81 **Topics in Current Chemistry**
Fortschritte der Chemischen Forschung

Large Amplitude Motion in Molecules I

Springer-Verlag Berlin Heidelberg GmbH 1979

This series presents critical reviews of the present position and future trends in modern chemical research. It is addressed to all research and industrial chemists who wish to keep abreast of advances in their subject.

As a rule, contributions are specially commissioned. The editors and publishers will, however, always be pleased to receive suggestions and supplementary information. Papers are accepted for "Topics in Current Chemistry" in English.

ISBN 978-3-662-15412-0 ISBN 978-3-540-35250-1 (eBook)
DOI 10.1007/978-3-540-35250-1

Library of Congress Cataloging in Publication Data. Main entry under title: Large amplitude motion in molecules I. (Topics in current chemistry ; 81) Bibliography: p. Includes index. 1. Molecular structure – Addresses, essays, lectures. I. Series. QDL.F58 · vol. 81 [QD461] · 540'.8s [541'.22] · 79-4221

Originally published by Springer-Verlag Berlin Heidelberg New York in 1979.
Softcover reprint of the hardcover 1st edition 1979

Triltsch, Graphischer Betrieb, 8700 Würzburg
2152/3140–543210

Contents

The Isometric Group of Nonrigid Molecules

Heinz Frei, Alfred Bauder and Hans H. Günthard

Laboratory for Physical Chemistry, ETHZ, CH-8092 Zürich, Switzerland

Table of Contents

1 Introduction

At the present time the conventional concept of molecular structure is appropriately based on the Born-Oppenheimer approximation[1]. Molecular structure is commonly understood as relative nuclear configuration, which may be considered as stable in the sense of one criterion or another. Many such structures may be characterized by a continuous set of nuclear configurations, which deviate only infinitesimally from each other (quasirigid molecules, sometimes called rigid molecules). Experimental research has revealed a large number of molecular structures which have to be described by a continuous set of nuclear configurations defined by structural parameters (bond length, bond angles, dihedral angles, etc.), some of which vary over finite domains. Molecules of this type will be called nonrigid molecules.

For quasirigid molecules a symmetry concept has been used very early in some branches of molecular research, e.g. stereochemistry[2, 3]. This symmetry concept was based on the concept of isometric mappings[4] and formed the basis of extended applications to molecular dynamics since 1930, developed first by Wigner[5].

Attempts to construct symmetries of nonrigid molecules have first been made by Hougen[6], Longuet-Higgins[7], and Altmann[8, 9]. All these procedures were based on the symmetries of the molecular Born-Oppenheimer operator, i.e. on the Schrödinger operator for a system of nuclei and electrons. In particular the Longuet-Higgins concept uses the intuitive concept of feasibility, which says that a permutation of nuclei corresponds to a feasible operation, if the permutation corresponds to a path on the Born-Oppenheimer surface involving only points of low potential energy. Hence, the elements of the Longuet-Higgins group are permutations and formal combinations of permutations and inversion. The whole concept lacks well defined mathematical tools for determination of transformation properties of energy operators, multipole operators and functions of the dynamical coordinates. Nevertheless, the concept has been applied to a number of specific examples, typical cases have been discussed by Hougen[10], but since its publication, the Longuet-Higgins concept has not been cast into a rigorous tool. Already before the Longuet-Higgins approach the symmetry of the rotation-internal motion problem of nonrigid molecules has been studied by direct investigation of the symmetry group of the rotation-internal motion hamiltonian. Typical examples of this direct approach have been given by Howard[11], Wilson[12], Wilson et al.[13].

The method presented here has been motivated by the desire to find a method which starts from the geometrical description of nuclear configurations and replaces the feasibility concept by rigorous mathematical definitions. Furthermore, it allows the determination of transformation properties of operators and functions by the methods used generally in applications of group theory to quantum mechanical problems in strict analogy to the treatment of quasirigid molecules within the framework of the covering symmetry group (molecular point symmetry group).

The approach presented in this contribution is a review of a method published in papers by Bauder et al.[14] and Frei et al.[15, 16]. It is based on the concept of the isometry of nuclear configurations and therefore may be considered as a natural generalization of the concept of covering symmetry of rigid point sets to nonrigid point sets.

In Chap. 2 the construction of the isometric group of semirigid nuclear configurations is presented, starting from the geometrical definitions of a semirigid model. Furthermore, the relation between isometric groups and the permutation-inversion group will be discussed. A number of applications of isometric groups, in particular to the dynamics of the rotation-large amplitude internal motion problem in classical and quantum mechanical formulation, to transformation properties of irreducible tensor operators and selection rules for irreducible tensor operators up to rank 2 (Wigner-Eckart theorem) are discussed in Chap. 3. Use of the isometric group to stereochemical problems of nonrigid molecules is presented, in particular for questions of chirality and classification of stereoisomers. In Chap. 4 relations of the isometric groups of semirigid models to the familiar symmetry approach for quasirigid molecules and to the symmetry groups of the associated nonrigid molecules are discussed.

For Chaps. 2 and 3 a number of examples will be given. Furthermore, techniques used for practical calculation of isometric groups and their application to problems of molecular geometry and dynamics will be collected in a series of appendices.

2 Construction of Isometric Groups

2.1 Definitions

By a nuclear configuration (NC) we understand the set of informations NC $\{\mathbf{X}_k, Z_k, M_k\}$ consisting of the coordinates $\mathbf{X}_k$, the masses M_k and charge numbers Z_k of the nuclei 1, 2, . . . , K of a molecular system. The coordinate vectors will be referred to a coordinate system, which will be defined when required. Important coordinate systems will be the laboratory system (LS, basis $\tilde{\mathbf{e}}^{l}$) and the frame system (FS, basis $\tilde{\mathbf{e}}^{f}$). The latter is attached to the nuclear configuration by a prescription to be defined in each case. The relation between $\tilde{\mathbf{e}}^{l}$ and $\tilde{\mathbf{e}}^{f}$ may be expressed by

$$\{\tilde{\mathbf{e}}^{f}\mathbf{X}^{f}\} = \{\tilde{\mathbf{e}}^{l}\mathbf{O}\} \cdot \begin{bmatrix} D(\epsilon) & X^{f} \\ 0 & 1 \end{bmatrix} \tag{2.1}$$

where $D(\epsilon) = \tilde{R}(\epsilon)$ is a rotation matrix parametrized by the eulerian angles $\alpha\beta\gamma$ (abbreviated by ϵ), as defined in Appendix 1. $\mathbf{X}^{f}$ stands for the origin of the FS with respect to the LS. For the dynamical problem $\mathbf{X}^{f}$ will be chosen as center-of-mass coordinate of the NC.

The relative nuclear configuration RNC $\{\mathbf{X}_k(\xi), Z_k, M_k\}$ is defined as the set of informations determining a NC up to translations and rotations in $\mathscr{R}_3$, i.e. invariant with respect to transformations of the inhomogeneous three-dimensional rotation group IO(3). Conveniently the RNC is determined by internal structural parameters $\xi_1, \xi_2, \ldots, \xi_{3K-6}$ which are invariant with respect to (w.r.t.) IO(3).

A molecule will be called rigid (quasirigid) if its internal structural parameters are constant (may vary only infinitesimally). The term semirigid model (SRM) will be used for a molecular model, whose nuclear configurations are defined by

1, 2, . ., $f \leqslant 3K\text{-}6$ internal coordinates which vary over finite domains, whereas the remaining 3K-6-f coordinates remain constant. The introduction of the SRM is motivated by the fact that its isometric group is isomorphic to the symmetry group of the associated nonrigid molecule (NRM), i.e. to the molecule with the same f finite and 3K-6-f infinitesimal internal coordinates (cf. Chap. 4). In practical cases the number of finite internal coordinates does not exceed 3 or 4 and remains always small in comparison to 3K-6.

If a NC of a SRM is considered from a suitably defined FS the coordinate vectors may be expressed as functions of the internal coordinates $\xi_1, \xi_2, \ldots, \xi_f$. The RNC $\{\mathbf{X}_k(\xi), Z_k, M_k\}$ is then completely defined by the values of $\xi_1, \ldots, \xi_f$ and the constant structural parameters. Further classification of SRMs may be based on the 2, 3, . . . rigid parts, whose relative positions are determined by the finite coordinates $\xi_1, \xi_2, \ldots$. Such parts are often denoted as frame (F), top (T), invertor (I), etc. Moreover each part may have its own local covering symmetry and the complete NC $\{\mathbf{X}_k(\xi), Z_k, M_k\}$ may have a proper covering symmetry group $\mathscr{G}(\xi)$ for arbitrary values of the internal coordinates. Typical SRMs are listed in Tables 1, 2 and 3.

To each NC we associate a graph $\mathscr{N}\,\{P(\pi_k(Z_k, M_k)), K(d_{kk'})\}$, consisting of the set P of vertices π_k valued by charge and mass number of the nucleus k and the set K of edges $(\pi_k, \pi_{k'})$, valued by the internuclear distance $d_{kk'}(\xi)$

$$d_{kk'}(\xi) = |\mathbf{X}_k(\xi) - \mathbf{X}_{k'}(\xi)| \tag{2.2}$$

$\mathscr{N}$ is a complete (universal) valued graph. In many cases it is sufficient to consider the graph $\underline{\mathscr{N}}\,\{P(\pi_k(Z_k)), K(d_{kk'})\}$ in which the vertices π_k are valued by the nuclear charge only. This is appropriate in all cases in which isotope effects within the Born-Oppenheimer approximation may be neglected.

2.2 Isometric Group of a SRM

The isometric group of a SRM will be constructed from two subgroups:
(i) internal isometric group $\mathscr{F}(\xi)$
(ii) covering group $\mathscr{G}(\xi)$

Since most of the nonrigid molecules treated so far may be described by a SRM whose covering group is the improper group C_1, the internal isometric group is treated first.

2.2.1 Internal Isometric Group $\mathscr{F}(\xi)$

From the definition (2.2) it is seen that the distances $d_{kk'}$ are functions of the internal coordinates. The set of transformations

$$\xi' = F(\xi) \tag{2.3}$$

which map the graph $\mathscr{N}$ onto itself, conserving incidence, forms a group $\mathscr{F}(\xi)$, the group law being the usual composition of functions. Mappings of the graph $\mathscr{N}$ onto itself are defined as

Table 1. Isometric groups of semirigid models $f = 1^a$, $\mathscr{G}(\xi) = C_1{}^b$

System	Type of int. motion	Covering symmetry		Symmetry groups				Typical molecule
		Frame	Top/invertor	$\Gamma^{(3)}\{\mathscr{S}\}$	$\mathscr{H}(\xi)$	$\overline{\mathscr{H}}(\xi)$	$\mathscr{I}\{H\}$	
$C_sF-C_{3v}T$	Int. rotation	C_s	C_{3v}	C_s	ϑ_3	ϑ_3	ϑ_3	CH_3CHO
$C_sF-C_{2v}T$	Int. rotation	C_s	C_{2v}	C_s	$\mathscr{V}_4$	$\mathscr{V}_4$	$\mathscr{V}_4$	$CH_2{:}CHNO_2$
C_sF-C_sT	Int. rotation	C_s	C_s	C_s	$\mathscr{V}_2$	$\mathscr{V}_2$	$\mathscr{V}_2$	$CH_2{:}CHCHO$
$C_{2v}F-C_{3v}T$	Int. rotation	C_{2v}	C_{3v}	C_{2v}	ϑ_6	ϑ_6	$\vartheta_6[E, T]^c$	CH_3NO_2
C_sF-C_sI	Inversion	C_s	C_s	C_s	$\mathscr{V}_2$	$\mathscr{V}_2$	$\mathscr{V}_2$	CHONHD
$C_sF-C_{2v}I$	Inversion	C_s	C_{2v}	C_s	$\mathscr{V}_2$	$\mathscr{V}_2$	$\mathscr{V}_2$	CH_2CR_2NH

a Number of finite degrees of freedom.
b Covering group for arbitrary values of ξ.
c T commutes with all other elements.

Table 2. Isometric groups of semirigid models $f \geqslant 2^a$, $\mathscr{G}(\xi) = C_1$[b]

f[a]	System	Type of motion[c]	Covering symmetry: Frame	Top 1	Top 2/ invertor	$\Gamma^{(3)}\{\mathscr{L}\}$	Symmetry groups: $\mathscr{H}(\xi)$	$\overline{\mathscr{H}}(\xi)$	Typical molecule
2	$C_{2v}F(C_{3v}T)_2$	RIR	C_{2v}	C_{3v}	C_{3v}	C_{2v}	$\mathscr{G}_{36}$	$\mathscr{G}_{36}$	CH_3COCH_3
2	$C_sF(C_{3v}T)(C_{3v}T)'$	RIR	C_s	C_{3v}	C_{3v}	C_s	$\mathscr{G}_{18}$	$\mathscr{G}_{18}$	$CH_3CH:NCH_3$
2	$C_sF(C_{3v}T)(C_{2v}T)$	RIR	C_s	C_{3v}	C_{2v}	C_s	ϑ_6	ϑ_6	$CH_3CH_2NO_2$
2	$C_sF(C_{3v}T)(C_sT)$	RIR	C_s	C_{3v}	C_s	C_s	ϑ_3	ϑ_3	CH_3CH_2CHO
2	$C_{2v}F(C_{2v}T)_2$	RIR	C_{2v}	C_{2v}	C_{2v}	C_{2v}	$\mathscr{G}_{16} = \vartheta_4[E, T]$[d]	$\mathscr{G}_{16}$	$CH_2(NO_2)_2$
2	$C_sF(C_{2v}T)(C_{2v}T)'$	RIR	C_s	C_{2v}	C_{2v}	C_s	$\mathscr{A}(2, 2, 2)$	$\mathscr{A}(2, 2, 2)$	$NO_2CH:CFNO_2$
2	$C_sF(C_{2v}T)(C_sT)$	RIR	C_s	C_{2v}	C_s	C_s	$\mathscr{V}_4$	$\mathscr{V}_4$	$NO_2CH_2CH_2F$
2	$C_{2v}F(C_sT)_2$	RIR	C_{2v}	C_s	C_s	C_{2v}	$\mathscr{V}_4$	$\mathscr{V}_4$	$O(CHO)_2$
2	$C_sF(C_sT)(C_sT)'$	RIR	C_s	C_s	C_s	C_s	$\mathscr{V}_2$	$\mathscr{V}_2$	CH_2FCH_2CHO
2	$C_{2v}F(C_{3v}T)(C_{2v}I)$	RIRINV	C_{2v}	C_{3v}	C_{2v}	C_{2v}	ϑ_6	ϑ_6	CH_3NH_2
3	$C_2(\tau)F(C_sT)_2$	RIR	$C_2(\tau)$	C_s	C_s	$C_{2v}(C_{2h})$	$\mathscr{V}_4$	$\mathscr{A}(2, 2, 2)$	CH_2OHCH_2OH

[a] Number of finite degrees of freedom.
[b] Covering group for arbitrary values of ξ.
[c] RIR = rotation-internal rotation, INV = inversion.
[d] T commutes with all other elements.

Table 3. Isometric Groups of Semirigid Models f = 1[a], $\mathscr{G}(\xi)$[b] proper

System	Type of int. motion	Symmetry groups						Fixed points	Typical molecule
		$\mathscr{F}(\xi)$	$\overline{\mathscr{F}}(\xi)$	$\mathscr{G}(\xi)$	$\mathscr{H}(\xi)$	$\overline{\mathscr{W}}(\xi)$	$\Gamma^{(3)}\{\overline{\mathscr{L}}\}$		
$D_{\infty h}F(C_{2v}T)_2$	Int. rot.	$\vartheta_2 \overset{is}{=} \mathscr{V}_4$	ϑ_4	D_2	$\vartheta_4\,[E, T]^c$	$\mathscr{G}_{32}$	D_{4h}	D_{2h}, D_{2d}	$(C_6H_5)_2$
$C_{2v}F-C_{2v}T$	Int. rot.	$\vartheta_2 \overset{is}{=} \mathscr{V}_4$	ϑ_2	C_2	$\mathscr{A}(2, 2, 2)$	$\mathscr{A}(2, 2, 2)$	C_{2v}	C_{2v}	$C_6H_5NO_2$
$D_{\infty h}F(C_sT)_2$	Int. rot.	$\mathscr{V}_2$	$\mathscr{V}_4$	C_2	$\mathscr{V}_4$	$\mathscr{A}(2, 2, 2)$	D_{2h}	C_{2v}, C_{2h}	CHOCHO
$C_{2v}F-C_{2v}I$	Inversion	$\mathscr{V}_2$	$\mathscr{V}_2$	C_s	$\mathscr{V}_4$	$\mathscr{V}_4$	C_{2v}	C_{2v}	$\overline{CH_2CH_2NH}$
$C_{3v}I$	Inversion	$\mathscr{V}_2$	$\mathscr{V}_4$	C_{3v}	ϑ_6	$\vartheta_6[E, T]^c$	D_{6h}	D_{3h}	NH_3

a Number of finite degrees of freedom.
b Covering group for arbitrary values of ξ.
c T commutes with all other elements.

$$\hat{F}: P \to P, K \to K$$

$$\hat{F}(\pi_k(Z_k, M_k)) = \pi_{\bar{k}}(Z_{\bar{k}}, M_{\bar{k}}) \in P, \quad k, \bar{k} \in [1, K] \tag{2.4}$$

where $Z_{\bar{k}} = Z_k, M_{\bar{k}} = M_k$

$$\hat{F}(d_{kk'}(\xi)) = d_{\bar{k}\bar{k}'}(\xi) \in K(d_{kk'}), \forall\, d_{kk'}(\xi) \tag{2.5}$$

The transformations $\xi' = F(\xi)$ will be called internal isometric transformations. They transform any NC to a NC with the same set of distances. In many cases they may be expressed as linear inhomogeneous transformations

$$\begin{pmatrix} \xi' \\ 1 \end{pmatrix} = \begin{pmatrix} A(F) & a(F) \\ 0 & 1 \end{pmatrix} \begin{pmatrix} \xi \\ 1 \end{pmatrix} = \mathscr{A}(F) \begin{pmatrix} \xi \\ 1 \end{pmatrix} \tag{2.6}$$

To the isometric transformation (2.3) we will associate the operator $\hat{P}_F$, defined by

$$\hat{P}_F h(\xi) = h(F^{-1}(\xi)) \tag{2.7}$$

where $h(\xi)$ is any admissible function of ξ. Application of $\hat{P}_F$ to the substrate $\widetilde{\{d_{kk'}(\xi)\}}$, i.e. to the set of distances $d_{kk'}$ ordered in a row yields

$$\hat{P}_F \widetilde{\{d_{kk'}(\xi)\}} = \widetilde{\{d_{kk'}(F^{-1}(\xi))\}} = \widetilde{\{d_{kk'}(\xi)\}}\, \Gamma^{(\mathscr{N}\mathscr{C})}(F) \tag{2.8}$$

The last equation expresses the fact that the set of distances is mapped by $\hat{P}_F$ onto itself, therefore the matrix $\Gamma^{(\mathscr{N}\mathscr{C})}(F)$ is a permutation matrix of dimension $\binom{K}{2}$, i.g. intransitive. The matrix group

$$\Gamma^{(\mathscr{N}\mathscr{C})}\{\mathscr{F}\} := \{\Gamma^{(\mathscr{N}\mathscr{C})}(F) \mid E, F_2, F_3, \ldots\} \tag{2.9}$$

is a representation of the isometric substitutions $\xi' = F(\xi)$ by permutation matrices. The symbol $\mathscr{F}(\xi)$ will henceforward be used as the abstract group $\mathscr{F}(\xi) :=$ $\{E, F_2, \ldots\}$ represented either by

$$\mathscr{A}\{\mathscr{F}\} := \left\{ \begin{pmatrix} A(F) & a(F) \\ 0 & 1 \end{pmatrix} \middle| \; \forall F \in \mathscr{F}(\xi) \right\}$$

$$\text{or by } \Gamma^{(\mathscr{N}\mathscr{C})}\{\mathscr{F}\} := \{\Gamma^{(\mathscr{N}\mathscr{C})}(F) \mid \forall F \in \mathscr{F}(\xi)\} \tag{2.10}$$

If the distances $d_{kk'}(\xi) \in K(d_{kk'})$ possess a common primitive period p w.r.t. the internal coordinates

$$d_{kk'}(\xi + p) = d_{kk'}(\xi), \quad \forall k, k' \in [1, K], \tag{2.11}$$

the coordinates involved in the transformations (2.6) have to be taken modulo their respective primitive periods. The implication of the existence of primitive periods will be discussed in Sect. 2.2.2.

The operators $\hat{P}_F$, $F \in \mathscr{F}(\xi)$ will next be applied to the basis $\widetilde{\{X_k(\xi)\}}$, i.e. to the (transposed) coordinate vectors referred to the frame system $\widetilde{e}^f$ ordered in a row:

$$\hat{P}_F\{\tilde{X}_k(\xi)\} = \{\tilde{X}_k(F^{-1}(\xi))\} = \{\tilde{X}_k(\xi)\}\, \Pi(F) \otimes \Gamma^{(3)}(F)$$
$$= \{\tilde{X}_k(\xi)\} \cdot \Gamma^{(NCf)}(F) \tag{2.12}$$

Thereby the matrix $\Pi(F)$ denotes a K-dimensional permutation matrix and $\Gamma^{(3)}(F)$ a 3 by 3 orthogonal matrix. The form of this representation follows from the fact that each isometric transformation maps the NC $\{\mathbf{X}_k, Z_k, M_k\}$ onto a NC which by definition has the same set of distances, i.e. is isometric to NC $\{X_k, Z_k, M_k\}$. Expressed alternatively, the nuclear configurations NC $\{\mathbf{X}_k(\xi), Z_k, M_k\}$ and NC $\{\mathbf{X}_k(F^{-1}(\xi)), Z_k, M_k\}$ are properly or improperly congruent up to permutations of nuclei with equal charge and mass for any $F \in \mathcal{F}(\xi)$. The set of matrices Eq. (2.12) forms a representation of $\mathcal{F}(\xi)$ by linear transformations and will furtheron be denoted by

$$\Gamma^{(NCf)}\{\mathcal{F}\} := \{\Pi(F) \otimes \Gamma^{(3)}(F) \mid \forall\, F \in \mathcal{F}(\xi)\} \tag{2.13}$$

the index f indicating reference to the frame system. In general $\Gamma^{(NCf)}\{\mathcal{F}\}$ decomposes into transitive systems, since each subset of identical nuclei, which is mapped by all elements of $\mathcal{F}$ onto itself gives rise to such a system. The group theoretical relation between $\mathcal{A}\ \{\mathcal{F}\}$ and $\Gamma^{(NCf)}\{\mathcal{F}\}$ is an isomorphism

$$\mathcal{A}\{\mathcal{F}\} \stackrel{\text{is}}{=} \Gamma^{(NCf)}\{\mathcal{F}\} \tag{2.14}$$

The isomorphism strictly holds for SRMs without primitive period isometric transformations only (cf. Sect. 2.2.2). However, as will be shown in Sect. 2.2.2, the group theoretical relations derived in this section also apply for SRMs with primitive period transformations if $\mathcal{F}$ is replaced by an appropriately extended group $\overline{\mathcal{F}}$. The sets

$$\Pi\{\mathcal{F}\} := \{\Pi(F) \mid \forall\, F \in \mathcal{F}(\xi)\} \tag{2.15}$$

$$\text{and } \Gamma^{(3)}\{\mathcal{K}\} := \{\Gamma^{(3)}(F) \mid \forall\, F \in \mathcal{F}(\xi)\} \tag{2.16}$$

form each a representation of $\mathcal{F}(\xi)$. The first set consisting of all permutation factors of $\Gamma^{(NCf)}\{\mathcal{F}\}$ is isomorphic to the permutation group $\Gamma^{(\mathcal{N}\mathcal{C})}\{\mathcal{F}\}$; this follows from a theorem given by Harary[17)], relating vertex and edge group of a complete graph.

The group $\Gamma^{(3)}\{\mathcal{K}\}$ (abstract group $\mathcal{K}$), consisting of all different rotational parts of $\Gamma^{(NCf)}$ is a finite group of orthogonal matrices in $\mathcal{R}_3$ and must be a subgroup of O(3). It therefore must be one of the point symmetry groups C_n, S_n, D_n, C_{nv}, C_{nh}, D_{nh}, D_{nd}, T, T_d, T_h, O, O_h. $\Gamma^{(3)}\{\mathcal{K}\}$ will play an important role in most applications of isometric groups. It pictures the set of all orthogonal matrices, which map a reference NC on to all possible isometric NCs.

In general the group theoretical relation between $\Gamma^{(NCf)}\{\mathcal{F}\}$ and $\Gamma^{(3)}\{\mathcal{K}\}$ is a homomorphism $\hat{\eta}$:

$$\hat{\eta} : \Gamma^{(NCf)}\{\mathcal{F}\} \longrightarrow \Gamma^{(3)}\{\mathcal{K}\} \tag{2.17}$$

whose kernel is given by

$$\ker \hat{\eta} := \{\Pi(F) \otimes 1^{(3)} \mid F \in \mathscr{F}(\xi)\} \qquad (2.17')$$

$$\frac{\Gamma^{(NCf)}\{\mathscr{F}\}}{\ker \hat{\eta}} \overset{\text{is}}{=} \Gamma^{(3)}\{\mathscr{K}\} \qquad (2.17'')$$

W.r.t. the structure of $\Gamma^{(3)}\{\mathscr{K}\}$ the following two cases will prove important:

Case a

$$\Gamma^{(3)}\{\mathscr{K}\} \text{ is properly orthogonal}, |\Gamma^{(3)}(K)| = 1, \forall K \in \mathscr{K} \qquad (2.18)$$

Case b

$\Gamma^{(3)}\{\mathscr{K}\}$ is improperly orthogonal, then one may write

$$\Gamma^{(3)}\{\mathscr{K}\} = \Gamma^{(3)}\{\mathscr{K}^+\} \cup \Gamma^{(3)}(T)\Gamma^{(3)}\{\mathscr{K}^+\} \qquad (2.19)$$

where $\Gamma^{(3)}\{\mathscr{K}^+\} := \{\Gamma^{(3)}(K) \mid K \in \mathscr{K}, |\Gamma^{(3)}(K)| = +1\}$ (2.20)

and $|\Gamma^{(3)}(T)| = -1$

Therefore[1] $\Gamma^{(3)}\{\mathscr{K}\} \overset{\text{ho}}{=} \mathscr{V}_2$ (2.21)

These group theoretical relations will find numerous applications in Chap. 3 and in fact are important in all applications of isometric groups.

In the study of the dynamical problem of SRMs the transformations of eulerian angles induced by isometric transformations of the frame system will be required. This leads in a natural way from the group $\Gamma^{(3)}\{\mathscr{K}\}$ to the group $\Delta^{(3)}\{\mathscr{K}\}$, defined as follows:

Case a

$$\Delta^{(3)}\{\mathscr{K}\} \equiv \Gamma^{(3)}\{\mathscr{K}\} \qquad (2.22a)$$

Case b

We have to distinguish between the following possibilities

1) $\Gamma^{(3)}\{\mathscr{K}\} = \Gamma^{(3)}\{\mathscr{K}^+\} \cup Z \cdot \Gamma^{(3)}\{\mathscr{K}^+\}$

where $Z = -1^{(3)}$,

then $\Delta^{(3)}\{\mathscr{K}\} \equiv \Gamma^{(3)}\{\mathscr{K}^+\}$ (2.22b1)

2) $\Gamma^{(3)}\{\mathscr{K}\} = \Gamma^{(3)}\{\mathscr{K}^+\} \cup \Gamma^{(3)}(T)\Gamma^{(3)}\{\mathscr{K}^+\}$

but $Z \notin \Gamma^{(3)}(T)\Gamma^{(3)}\{\mathscr{K}^+\}$

then any element of the coset may be written as a product of Z with a properly orthogonal matrix. $\Gamma^{(3)}(T)$ may then be written as

---

1 $\mathscr{V}_2$ denotes the two-group.

$$\Gamma^{(3)}(T) = Z \cdot R,\ R \in SO(3),\ R \notin \Gamma^{(3)}\{\mathcal{K}^+\}$$

Hence

$$\Gamma^{(3)}\{\mathcal{K}\} = \Gamma^{(3)}\{\mathcal{K}^+\} \cup Z \cdot R \cdot \Gamma^{(3)}\{\mathcal{K}^+\}$$

and

$$\Delta^{(3)}\{\mathcal{K}\} = \Gamma^{(3)}\{\mathcal{K}^+\} \cup R \cdot \Gamma^{(3)}\{\mathcal{K}^+\} \overset{\text{is}}{=} \Gamma^{(3)}\{\mathcal{K}\} \tag{2.22b2}$$

$\Delta^{(3)}\{\mathcal{K}\}$ is always identical with a finite subgroup of SO(3).

In Fig. 1 the groups $\mathcal{A}\{\mathcal{F}\}$, $\Gamma^{(\mathcal{N}\mathcal{C})}\{\mathcal{F}\}$, $\Gamma^{(NCf)}\{\mathcal{F}\}$, $\Gamma^{(3)}\{\mathcal{K}\}$ and $\Delta^{(3)}\{\mathcal{K}\}$ are shown together with their group theoretical relations.

2.2.1.1 Transformation Group of the Dynamical Variables. The transformation groups $\Gamma^{(NCf)}\{\mathcal{F}\}$, $\Gamma^{(3)}\{\mathcal{K}\}$ and $\Delta^{(3)}\{\mathcal{K}\}$ all refer to the frame system $\tilde{e}^f$. By means of the relation between the frame and laboratory system Eq. (2.1) they may be used to define the transformations of the eulerian angles as follows:

1) with each transformation $\Gamma^{(3)}(F) \in \Gamma^{(3)}\{\mathcal{K}\}$ we may associate a basis transformation

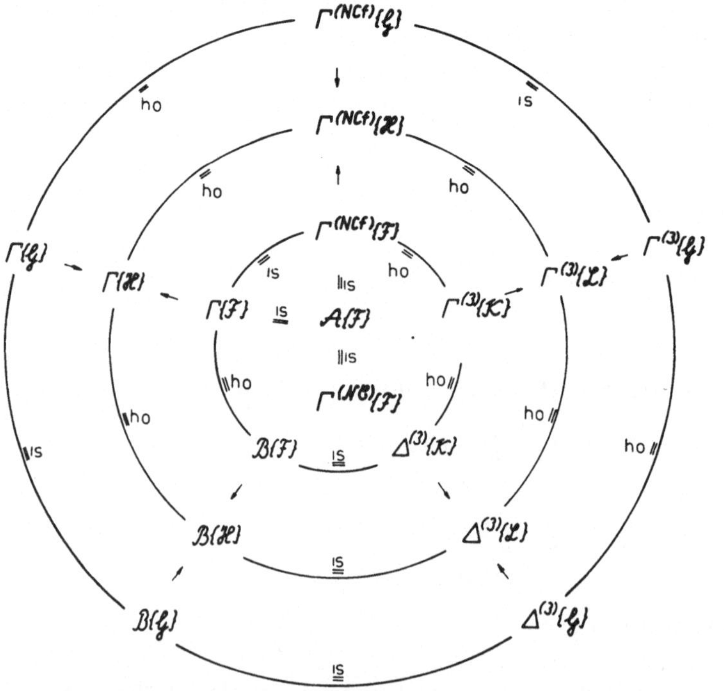

Fig. 1. Group theoretical interrelations between representations of the isometric group.
Key: For SRMs with primitive period isometric transformations the isomorphisms hold strictly for the representations of $\mathcal{F}$, $\mathcal{K}$, $\mathcal{H}$ and $\mathcal{G}$, but $\mathcal{A}\{\mathcal{F}\}$ is homomorphic onto $\Gamma^{(\mathcal{N}\mathcal{F})}\{\mathcal{F}\} \overset{\text{is}}{=} \Gamma^{(\mathcal{N}\mathcal{C})}\{\mathcal{F}\}$

$$\tilde{\mathbf{e}}^{f'} = \tilde{\mathbf{e}}^{f}\tilde{\Gamma}^{(3)}(F),\ F \in \mathscr{F} \qquad (2.23)$$

and the contragredient coordinate transformation

$$(X') = \Gamma^{(3)}(F)(X) \qquad (2.24)$$

2) accordingly one may associate the transformation

$$\tilde{\mathbf{e}}^{f'} = \tilde{\mathbf{e}}^{1}D(\epsilon)\tilde{\Gamma}^{(3)}(F) = \tilde{\mathbf{e}}^{1}D(\epsilon')\ |\Gamma^{(3)}(F)| \qquad (2.25)$$

which implies

$$D(\epsilon') = D(\epsilon)\tilde{\Gamma}^{(3)}(F)\ |\Gamma^{(3)}(F)|$$

or

$$D(\epsilon') = D(\epsilon)\tilde{\Gamma}^{(3)}(F) \text{ if } |\Gamma^{(3)}(F)| = +1 \qquad (2.26)$$

$$D(\epsilon') = D(\epsilon)\tilde{R}(F) \text{ if } |\Gamma^{(3)}(F)| = -1 \qquad (2.26')$$

where $R(F) = \Gamma^{(3)}(F)Z$

The Eqs. (2.26) define transformations of the eulerian angles

$$\epsilon' = \epsilon'(\epsilon, F) \qquad (2.27)$$

which in most practical cases are linear inhomogeneous

$$\begin{pmatrix}\epsilon'\\1\end{pmatrix} = \begin{pmatrix}B(F) & b(F)\\0 & 1\end{pmatrix}\begin{pmatrix}\epsilon\\1\end{pmatrix} = \mathscr{B}(F)\begin{pmatrix}\epsilon\\1\end{pmatrix} \qquad (2.28)$$

The proper set $\mathscr{B}\{\mathscr{F}\}$

$$\mathscr{B}\{\mathscr{F}\} := \{\mathscr{B}(F) \mid \forall F \in \mathscr{F}\} \qquad (2.29)$$

forms a group, which is isomorphic to $\Delta^{(3)}\{\mathscr{K}\}$

$$\mathscr{B}\{\mathscr{F}\} \overset{\text{is}}{=} \Delta^{(3)}\{\mathscr{K}\} \qquad (2.30)$$

From the Eqs. (2.17), (2.22) and (2.30) we obtain

$$\Gamma^{(NCf)}\{\mathscr{F}\} \overset{\text{ho}}{=} \mathscr{B}\{\mathscr{F}\} \qquad (2.31)$$

Next we consider the direct sum of the two transformation groups $\mathscr{A}\{\mathscr{F}\}$ and $\mathscr{B}\{\mathscr{F}\}$. In the case where both these groups may be represented by linear inhomogeneous transformations according to Eqs. (2.6), (2.28) this leads to the matrix group

$$\Gamma\{\mathscr{F}\} := \left\{ \Gamma(F) = \begin{bmatrix} B(F) & 0 & b(F) \\ 0 & A(F) & a(F) \\ 0 & 0 & 1 \end{bmatrix} \middle| \forall F \in \mathscr{F} \right\} \tag{2.32}$$

$$\begin{bmatrix} \epsilon' \\ \xi' \\ 1 \end{bmatrix} = \Gamma(F) \begin{bmatrix} \epsilon \\ \xi \\ 1 \end{bmatrix} \tag{2.33}$$

It follows directly from Eqs. (2.14) and (2.31) that

$$\mathscr{A}\{\mathscr{F}\} \overset{\text{ho}}{=} \mathscr{B}\{\mathscr{F}\} \tag{2.34}$$

Therefore,

$$\{\mathscr{B}(F) \oplus \mathscr{A}(F) \mid \forall F \in \mathscr{F}(\xi)\} = \Gamma\{\mathscr{F}\} \overset{\text{is}}{=} \mathscr{A}\{\mathscr{F}\} \tag{2.35}$$

and by Eq. (2.14)

$$\Gamma\{\mathscr{F}\} \overset{\text{is}}{=} \Gamma^{(\text{NCf})}\{\mathscr{F}\} \tag{2.36}$$

A very important relation is

$$\Gamma\{\mathscr{F}\} \overset{\text{ho}}{=} \Gamma^{(3)}\{\mathscr{K}\} \tag{2.37}$$

and, therefore, for case b SRMs

$$\Gamma\{\mathscr{F}\} \overset{\text{ho}}{=} \mathscr{V}_2 \tag{2.38}$$

These relations are symbolically represented in Fig. 1.

2.2.2 Primitive Period Isometric Transformations

The investigation of the set of distances $\{\widetilde{d_{kk'}(\xi)}\}$ w.r.t. isometric transformations in many cases leads to transformations of the type

$$\begin{pmatrix} \xi' \\ 1 \end{pmatrix} = \begin{pmatrix} 1^{(f)} & p \\ 0 & 1 \end{pmatrix} \begin{pmatrix} \xi \\ 1 \end{pmatrix} = \mathscr{A}(F_p) \begin{pmatrix} \xi \\ 1 \end{pmatrix} \tag{2.39}$$

$$\text{with } d_{kk'}(\xi + p) = d_{kk'}(\xi), \quad \forall k, k' \in [1, K] \tag{2.11}$$

but $\{\widetilde{X}_k(\xi + p)\} \neq \{\widetilde{X}_k(\xi)\}$. Thereby p denotes the primitive period of the distances. Isometric transformations of this type will hereafter be called primitive period isometric transformations. Equation (2.11) expresses that F_p maps the graph $\mathscr{N}$ identically onto itself[2]

---

[2] $1^{\binom{K}{2}}$ denotes a unit matrix of dimension $\binom{K}{2}$.

$$\Gamma^{(\mathscr{N}\mathscr{C})}(F_p) = 1^{\binom{K}{2}} \tag{2.40}$$

However, NC $\{X_k(\xi), Z_k, M_k\}$ and NC $\{X_k(\xi - p), Z_k, M_k\}$ are not identical, but may be mapped onto each other by a nontrivial element of O(3). Associating operators $\hat{P}_{F_p}$ with primitive period transformations of type (2.39) allows to express this relation by

$$\begin{aligned} \hat{P}_{F_p}\{\tilde{X}_k(\xi)\} = \{\tilde{X}_k(\xi - p)\} &= \{\tilde{X}_k(\xi)\}\, 1^{(K)} \otimes \Gamma^{(3)}(F_p) \\ &= \{\tilde{X}_k(\xi)\}\, \Gamma^{(NCf)}(F_p) \end{aligned} \tag{2.41}$$

where $\Gamma^{(3)}(F_p) \neq 1^{(3)}$, $\Gamma^{(3)}(F_p) \in O(3)$

Therefore,

$$\Gamma(F_p) = \begin{bmatrix} B(F_p) & 0 & b(F_p) \\ . & 1^{(f)} & p \\ . & . & 1 \end{bmatrix} \neq 1^{(f+4)} \tag{2.41'}$$

Whereas the group $\mathscr{F}$ and its representations are relevant and sufficient for problems which are completely defined by relative nuclear configurations (RNCs) of a SRM, primitive period isometric transformations have to be considered as nontrivial symmetry operations in all those applications where the orientation of the NC w.r.t. the frame and laboratory coordinate system is relevant, e.g. the rotation-internal motion energy eigenvalue problem of a SRM. Inclusion of such primitive period operations leads to the internal isometric group $\bar{\mathscr{F}}(\xi)$ represented faithfully by

$$\mathscr{A}\{\bar{\mathscr{F}}\} := \left\{ \begin{pmatrix} A(F) & a(F) \\ 0 & 1 \end{pmatrix} \mid \forall F \in \bar{\mathscr{F}}(\xi) \right\} \tag{2.42}$$

For SRMs with nontrivial primitive period transformations one has thus to distinguish between two types of internal isometric groups: $\mathscr{F}(\xi)$ and $\bar{\mathscr{F}}(\xi)$. The former group is defined as the abstract group of the representation $\Gamma^{(\mathscr{N}\mathscr{C})}\{\mathscr{F}\}$ on the distance set [Eq. (2.9)] and does not include primitive period transformations by virtue of Eq. (2.40). $\mathscr{F}$ is alternatively defined as the substitution group $\mathscr{A}\{\mathscr{F}\}$ [Eq. (2.10)] in which the internal coordinates are taken modulo p (cf. Sect. 2.2.1). For SRMs with primitive period transformations neither $\Gamma^{(NCf)}$ nor Γ necessarily contain subgroups which are isomorphic to $\mathscr{F}(\xi)$. However, the isomorphisms between $\mathscr{A}$ and $\Gamma^{(NCf)}$ and $\mathscr{A}\,\Gamma\{\mathscr{F}\}$ [Eqs. (2.14) and (2.35), Sect. 2.2.1] hold always for $\bar{\mathscr{F}}(\xi)$

$$\mathscr{A}\{\bar{\mathscr{F}}\} \stackrel{is}{=} \Gamma^{(NCf)}\{\bar{\mathscr{F}}\} \tag{2.14'}$$

$$\mathscr{A}\{\bar{\mathscr{F}}\} \stackrel{is}{=} \Gamma\{\bar{\mathscr{F}}\} \tag{2.35'}$$

The group $\mathscr{A}\{\bar{\mathscr{F}}\}$ generates the whole set of representations $\Gamma^{(NCf)}\{\bar{\mathscr{F}}\}$, $\Gamma^{(3)}\{\bar{\mathscr{K}}\}$ etc. according to the innermost circle of Fig. 1.

At the present time the following two cases concerning the group theoretical relation between $\mathscr{F}$ and $\overline{\mathscr{F}}$ have been observed
(1) $\overline{\mathscr{F}}$ is a cover group of $\mathscr{F}$, i.e.

$$\overline{\mathscr{F}} \overset{\text{ho}}{=} \mathscr{F} \tag{2.43}$$

(2) $\mathscr{F}$ is at the same time a subgroup of $\overline{\mathscr{F}}$, i.e.[3]

$$\mathscr{F} \subset \overline{\mathscr{F}}, \ \mathscr{F} \overset{\text{end}}{=} \overline{\mathscr{F}} \tag{2.43'}$$

It should be pointed out that the occurrence of primitive period transformations is closely connected to the choice of the internal coordinates or equivalently to the choice of the frame system. Expectation values of all observable quantities whether dependent only on the RNC or dependent on the NC of a SRM must be independent of the choice of internal coordinates (frame system). If introduction of a certain internal coordinate (frame system) gives rise to a primitive period transformation and as a consequence to an extension of $\mathscr{F}$ to $\overline{\mathscr{F}}$, then still observable quantities should be classifyable according to the symmetry group $\mathscr{F}$.

2.2.3 Covering Group $\mathscr{G}(\xi)$

SRMs often exhibit nontrivial covering symmetries (in the sense of covering symmetries of a rigid point set) for arbitrarily chosen but fixed values of the internal coordinates ξ. It is evident that such covering operations are isometric mappings of a point set onto itself and therefore have to be included in the full group of isometric transformations. The group of covering operations will be denoted by $\mathscr{G}(\xi)$. For the definition of the operators $\hat{P}_G$, $G \in \mathscr{G}(\xi)$, acting on the coordinate vectors we consider a subset of equivalent nuclei in general site w.r.t. $\mathscr{G}(\xi)$ whose coordinate vectors are generated from a representative nucleus by the mappings

$$(X_{G_k}) = \widetilde{\Gamma}^{(3)}(G_k)(X_E), \ G_k \in \mathscr{G}(\xi) \tag{2.44}$$

the nuclei being labeled here by the group elements G_k. The covering group $\mathscr{G}(\xi)$ is defined by the set of coordinate transformations

$$(X') = \Gamma^{(3)}(G)(X) \tag{2.45}$$

that means the abstract group $\mathscr{G}(\xi) = \{E, G_2, \ldots, G_{|\mathscr{G}|}\}$[4] is defined by the matrix group

$$\{\Gamma^{(3)}(G) \mid (X') = \Gamma^{(3)}(G)(X)\} \tag{2.45'}$$

---

3 $\overset{\text{end}}{=}$ Denotes endomorphic.

4 $|\mathscr{G}|$ Denotes the order of the group $\mathscr{G}(\xi)$.

The basis transformations associated to the coordinate transformations (2.45) are

$$\tilde{\mathbf{e}}^{f'} = \tilde{\mathbf{e}}^{f}\Gamma^{(3)}(G) \tag{2.46}$$

To the coordinate transformations (2.45) there are associated operators $\hat{P}_G$, defined by[18)]

$$\hat{P}_G f(X) = f(\Gamma^{(3)}(G)^{-1}(X)) \tag{2.47}$$

Using this definition we get for one vector

$$\hat{P}_G \tilde{X} = \widetilde{(\Gamma^{(3)}(G)^{-1}(X))} = \tilde{X}\Gamma^{(3)}(G) \tag{2.47'}$$

and, if $\hat{P}_G$ is applied to the set of coordinate vectors (2.44) arranged in a row

$$\{\tilde{X}_{G_k}(\xi)\} = \{\tilde{X}_E(\xi), \tilde{X}_E(\xi)\cdot\Gamma^{(3)}(G_2), \ldots\} \tag{2.44'}$$

we get

$$\hat{P}_G\{\tilde{X}_{G_k}(\xi)\} = \{\tilde{X}_{G_k}(\xi)\cdot\Gamma^{(3)}(G)\} = \{\tilde{X}_{G_k}(\xi)\}\,1^{|\mathscr{G}|}\otimes\Gamma^{(3)}(G) \tag{2.48}$$

The set of matrices

$$\{1^{|\mathscr{G}|}\otimes\Gamma^{(3)}(G)\,|\,\forall G\in\mathscr{G}(\xi)\} \overset{\text{is}}{=} \mathscr{G}(\xi) \tag{2.49}$$

forms a faithful representation of the covering group $\mathscr{G}(\xi)$. Because $\tilde{X}_{G_k}\cdot\Gamma^{(3)}(G)$ represents the vector $\underset{\sim}{X}_{G_kG}$ according to Eq. (2.44), the transformation (2.48) may equally well be expressed by a permutation

$$\hat{P}_G\{\tilde{X}_{G_k}(\xi)\} = \{\tilde{X}_{G_k}(\xi)\}\,1^{|\mathscr{G}|}\otimes\Gamma^{(3)}(G) = \{\tilde{X}_{G_k}(\xi)\}\,\Lambda(G)\otimes 1^{(3)} \tag{2.48'}$$

where $\Lambda(G)$ denotes a $|\mathscr{G}|$ dimensional permutation matrix. However, the matrix group

$$\{\Lambda(G)\otimes 1^{(3)}\,|\,\forall G\in\mathscr{G}(\xi)\} \tag{2.50}$$

is antiisomorphic[19)] to $\mathscr{G}(\xi)$ defined by the matrix group (2.45'), i.e.

$$\Lambda(G_2)\Lambda(G_1) = \Lambda(G_1G_2) \tag{2.51}$$

as may easily be proved by calculating $\hat{P}_{G_2}\hat{P}_{G_1}\{\tilde{X}_{G_k}(\xi)\}$[20)]. On the other hand the group of matrices

$$\{\Pi(G)\otimes 1^{(3)}\,|\,\forall G\in\mathscr{G}(\xi)\} \overset{\text{is}}{=} \mathscr{G}(\xi) \tag{2.49'}$$

$$\text{with } \Pi(G) = \Lambda(G^{-1}),\ \forall G\in\mathscr{G}(\xi) \tag{2.52}$$

forms a faithful representation of $\mathcal{G}(\xi)$ by permutations, as follows directly from Eq. (2.51)

$$\Lambda(G_2^{-1})\Lambda(G_1^{-1}) = \Lambda((G_2G_1)^{-1}) \tag{2.51'}$$

therefore $\Pi(G_2)\Pi(G_1) = \Pi(G_2G_1)$

For a set of equivalent nuclei in general site the matrices $\Pi(G)$ are identical with the right regular representation matrices[21]. If the nuclear position vectors of all K nuclei of a SRM are included in the basis $\{\widetilde{X}_k(\xi)\}$, $\Pi(G)$ denotes a K by K permutation matrix. In addition to the matrix groups (2.49) and (2.49′) the set

$$\Gamma^{(NCf)}\{\mathcal{G}\} := \{\Pi(G)\otimes\Gamma^{(3)}(G)\,|\,\forall G\in\mathcal{G}(\xi)\} \overset{\text{is}}{=} \mathcal{G}(\xi) \tag{2.49''}$$

forms a faithful representation of the covering group $\mathcal{G}(\xi)$ since both factors of the direct product form groups isomorphic to $\mathcal{G}(\xi)$. Thereby, all elements of $\Gamma^{(NCf)}\{\mathcal{G}\}$ map each coordinate vector identically onto itself, cf. Eq. (2.48′)

$$\begin{aligned}\hat{P}_G\hat{P}_G^{-1}\{\widetilde{X}_k(\xi)\} &= \{\widetilde{X}_k(\xi)\} = \{\widetilde{X}_k(\xi)\}\,\Lambda(G^{-1})\otimes\Gamma^{(3)}(G)\\ &= \{\widetilde{X}_k(\xi)\}\,\Pi(G)\otimes\Gamma^{(3)}(G),\;\; \forall G\in\mathcal{G}(\xi)\end{aligned} \tag{2.53}$$

The group (2.49″) will prove important for the construction of the full isometric group, cf. Sect. 2.2.4.

From the definition of covering symmetry which basically rests on the concept of the isometric mapping of a point set onto itself, it is evident that the operators $\hat{P}_G$ map the distance set $\{\widetilde{d_{kk'}(\xi)}\}$ onto itself by intransitive permutations:

$$\hat{P}_G\{\widetilde{d_{kk'}(\xi)}\} = \{\widetilde{d_{kk'}(\xi)}\}\,\Gamma^{(\mathcal{N}\mathcal{C})}(G) \tag{2.54}$$

The set $\Gamma^{(\mathcal{N}\mathcal{C})}\{\mathcal{G}\} := \{\Gamma^{(\mathcal{N}\mathcal{C})}(G)\,|\,\forall G\in\mathcal{G}(\xi)\}$ (2.55)

forms an intransitive representation of $\mathcal{G}(\xi)$ by permutations. In analogy to the relation between $\Gamma^{(\mathcal{N}\mathcal{C})}\{\mathcal{F}\}$ and $\Pi\{\mathcal{F}\}$ one has by the same argument (cf. Sect. 2.2.1)

$$\Gamma^{(\mathcal{N}\mathcal{C})}\{\mathcal{G}\} \overset{\text{is}}{=} \Pi\{\mathcal{G}\} \tag{2.56}$$

where the group $\Pi\{\mathcal{G}\}$ is defined by the set of matrices $\Pi(G)$ Eq. (2.49′).

Starting with the representation $\Gamma^{(NCf)}\{\mathcal{G}\}$ one may now construct representations of $\mathcal{G}(\xi)$ on the various substrates in strict analogy to the procedure applied for the internal isometric group $\mathcal{F}(\xi)$. The various steps of the construction are symbolized on the outermost circle of Fig. 1. This leads successively

(i) from $\Gamma^{(NCf)}\{\mathcal{G}\}$ to $\Gamma^{(3)}\{\mathcal{G}\}$:

$$\Gamma^{(3)}\{\mathcal{G}\} := \{\Gamma^{(3)}(G)\,|\,\forall G\in\mathcal{G}(\xi)\} \tag{2.57}$$

(ii) from $\Gamma^{(3)}\{\mathcal{G}\}$ to $\Delta^{(3)}\{\mathcal{G}\}$, where one has to distinguish between

Case a:

$\Gamma^{(3)}\{\mathcal{G}\}$ properly orthogonal, hence

$$\Gamma^{(3)}\{\mathcal{G}\} \equiv \Delta^{(3)}\{\mathcal{G}\} \tag{2.58}$$

Case b:

$\Gamma^{(3)}\{\mathcal{G}\}$ improperly orthogonal, hence

$\Gamma^{(3)}\{\mathcal{G}\} = \Gamma^{(3)}\{\mathcal{G}^+\} \cup \Gamma^{(3)}(T)\Gamma^{(3)}\{\mathcal{G}^+\}$ where $|\Gamma^{(3)}(T)| = -1$. In this case

$$\Gamma^{(3)}\{\mathcal{G}\} \overset{\text{ho}}{=} \mathcal{V}_2 \tag{2.59}$$

There exists an analogous differentiation of subcases b1 and b2, as has been discussed in Sect. 2.2.1, Eqs. (2.22b).

(iii) from $\Delta^{(3)}\{\mathcal{G}\}$ to $\mathcal{B}\{\mathcal{G}\}$ (in analogy to the transition from $\Delta^{(3)}\{\mathcal{K}\}$ to $\mathcal{B}\{\mathcal{F}\}$):

$$\mathcal{B}\{\mathcal{G}\} := \left\{ \begin{pmatrix} \epsilon' \\ 1 \end{pmatrix} = \begin{pmatrix} B(G) & b(G) \\ 0 & 1 \end{pmatrix} \begin{pmatrix} \epsilon \\ 1 \end{pmatrix} \middle| \forall G \in \mathcal{G}(\xi) \right\} \tag{2.60}$$

where the transformation of the eulerian angles ϵ are induced by the basis transformation Eq. (2.46) defining the covering operation and the relation

$$\tilde{e}^{f'} = \tilde{e}^1 D(\epsilon)\tilde{\Gamma}^{(3)}(G) = \tilde{e}^1 D(\epsilon')|\Gamma^{(3)}(G)| \tag{2.61}$$

(iv) from $\mathcal{B}\{\mathcal{G}\}$ and $\mathcal{A}\{\mathcal{G}\}$ to $\Gamma\{\mathcal{G}\}$:

$$\Gamma\{\mathcal{G}\} := \left\{ \begin{bmatrix} \epsilon' \\ \xi' \\ 1 \end{bmatrix} = \begin{bmatrix} B(G) & 0 & b(G) \\ . & 1^{(f)} & 0 \\ . & . & 1 \end{bmatrix} \cdot \begin{bmatrix} \epsilon \\ \xi \\ 1 \end{bmatrix} \middle| \forall G \in \mathcal{G}(\xi) \right\} \tag{2.62}$$

since by definition

$$\mathcal{A}(G) = \begin{pmatrix} 1^{(f)} & 0 \\ 0 & 1 \end{pmatrix}, \quad \forall G \in \mathcal{G}(\xi) \tag{2.63}$$

in agreement with the conventions proposed above w.r.t. $\mathcal{G}(\xi)$. Again the group $\Gamma\{\mathcal{G}\}$ will prove important in the consideration of the rotation-internal motion problem of SRMs.

Typical examples for SRMs with proper covering group are listed in Table 3.

2.2.3.1 Fixed Points of Isometric Transformations. In many SRMs special values of the internal coordinates ξ occur, which define NC$\{\mathbf{X}_k, \mathbf{Z}_k, \mathbf{M}_k\}$ of higher covering symmetry than $\mathcal{G}(\xi)$; if ξ_F is such a point in the parameter space, then

$$\mathcal{G}(\xi) \subset \mathcal{G}(\xi_F) \tag{2.64}$$

The relation of such special points to internal isometric transformations may be derived as follows.

Assume the internal isometric transformation

$$\begin{pmatrix}\xi'\\1\end{pmatrix} = \begin{pmatrix}A(F) & a(F)\\0 & 1\end{pmatrix}\begin{pmatrix}\xi\\1\end{pmatrix}$$

to have a fixed point, i.e. the equation

$$\begin{pmatrix}\xi\\1\end{pmatrix} = \begin{pmatrix}A(F) & a(F)\\0 & 1\end{pmatrix}\begin{pmatrix}\xi\\1\end{pmatrix}$$

or

$$(A(F) - 1^{(f)})\,\xi + a(F) = 0 \tag{2.65}$$

has a solution ξ_F. As a consequence

$$\begin{aligned}\hat{P}_F\{\tilde{X}_k(\xi_F)\} = \{\tilde{X}_k(\xi_F)\} = \{\tilde{X}_k(\xi_F)\}\,\Pi(F)\otimes\Gamma^{(3)}(F),\\ \{\tilde{X}_k(\xi_F)\}\,\Pi(F^{-1})\otimes 1^{(3)} = \{\tilde{X}_k(\xi_F)\}\,1^{(K)}\otimes\Gamma^{(3)}(F)\end{aligned} \tag{2.66}$$

The last Eq. (2.66) implies that the mapping on the right hand side generates merely a permutation of the coordinate vectors, i.e. $\Gamma^{(3)}(F)$ must be a covering operation of the NC $\{X_k(\xi_F), Z_k, M_k\}$. One therefore may state that every fixed point is connected to a covering symmetry operation not contained in $\mathscr{G}(\xi)$. If $F \in \mathscr{F}(\xi)$ is an isometric transformation with fixed point, then also the period of F

$$\mathscr{P}\{F\} := \{F^k, k = 1, 2, \ldots\} \tag{2.67}$$

has the fixed point ξ_F. Furthermore, if two different isometric transformations have a common fixed point, then the whole group $\mathscr{C}\{F_i, F_k\}$ generated by the two transformations has the same fixed point. Any such group $\mathscr{C}$ must be a subgroup of $\mathscr{F}(\xi)$:

$$\mathscr{C}\{F_i, F_k\} \subseteq \mathscr{F}(\xi) \tag{2.68}$$

It should be pointed out that the fixed points play an important role in geometrical application of isometric groups, e.g. stereochemistry, cf. Sects. 3.4 and 3.5.

2.2.4 Full Isometric Group $\mathscr{H}(\xi)$

The two groups $\mathscr{F}(\xi)$ and $\mathscr{G}(\xi)$ generate an abstract group $\mathscr{H}(\xi)$, whose representations $\Gamma^{(\mathscr{N}\mathscr{C})}\{\mathscr{H}\}$, $\Gamma^{(NCf)}\{\mathscr{H}\}$, $\Gamma^{(3)}\{\mathscr{L}\}$, ... are generated by the pairs of corresponding representations of $\mathscr{F}$ and $\mathscr{G}$ indicated on the innermost and outermost circles of the group diagram Fig. 1. The diagram shows the relationship of the various representations of $\mathscr{H}(\xi)$.

In Sect. 2.2.2 we have shown that if a SRM admits primitive period isometric transformations, representations of two groups $\mathscr{F}$ and $\overline{\mathscr{F}}$ may be derived. Extension of a representation of $\mathscr{F}$ by $\mathscr{G}$ leads to the corresponding representation of $\mathscr{H}$, whereas extension of the representations of $\overline{\mathscr{F}}$ by $\mathscr{G}$ gives those of $\overline{\mathscr{H}}$. The use of $\mathscr{H}$ or $\overline{\mathscr{H}}$ depends on the problem to which the isometric group is to be applied, as has been pointed out in Section 2.2.2. In order to simplify the notation we shall for general discussions not distinguish between the representations of $\mathscr{H}$ and $\overline{\mathscr{H}}$.

The structure of the group $\mathscr{H}(\xi)$ may be derived by considering the solutions of Eq. (2.12). This equation admits $|\mathscr{G}|$ solutions for each $F \in \mathscr{F}(\xi)$ which by aid of Eq. (2.53) may be written as

$$\begin{aligned} \hat{P}_F\{\widetilde{X}_k(\xi)\} = \hat{P}_G\hat{P}_G^{-1}\hat{P}_F\{\widetilde{X}_k(\xi)\} &= \{\widetilde{X}_k(\xi)\}\Lambda(G^{-1})\Pi(F) \otimes \Gamma^{(3)}(G)\Gamma^{(3)}(F) \\ &= \{\widetilde{X}_k(\xi)\}\Pi(G)\Pi(F) \otimes \Gamma^{(3)}(G)\Gamma^{(3)}(F) \\ &\forall G \in \mathscr{G}(\xi) \end{aligned} \tag{2.69}$$

Therefore, the internal isometric group $\mathscr{F}(\xi)$ generates on the position vectors of the nuclei a set of $|\mathscr{G}| \cdot |\mathscr{F}|$ matrices

$$\{\Pi(G)\Pi(F) \otimes \Gamma^{(3)}(G)\Gamma^{(3)}(F) \mid \forall G \in \mathscr{G}(\xi), \forall F \in \mathscr{F}(\xi)\} \tag{2.70}$$

For $E \in \mathscr{F}(\xi)$ we have

$$\hat{P}_E\{\widetilde{X}_k(\xi)\} = \hat{P}_G\hat{P}_G^{-1}\{\widetilde{X}_k(\xi)\} = \{\widetilde{X}_k(\xi)\} = \{\widetilde{X}_k(\xi)\}\Pi(G) \otimes \Gamma^{(3)}(G),$$
$\forall G \in \mathscr{G}(\xi)$, cf. Eq. (2.53)

The set of matrices $\Pi(G) \otimes \Gamma^{(3)}(G)$, i.e. the group $\Gamma^{(NCf)}\{\mathscr{G}\}$ Eq. (2.49″) forms an invariant subgroup of the group generated by the set (2.70). This follows directly from the definition

$$\hat{P}_F\hat{P}_G\hat{P}_G^{-1}\hat{P}_F^{-1} = \hat{P}_E \tag{2.71}$$

and is verified explicitely by the equation

$$\begin{aligned} \hat{P}_F\hat{P}_G\hat{P}_G^{-1}\hat{P}_F^{-1}\{\widetilde{X}_k(\xi)\} &= \{\widetilde{X}_k(\xi)\}\Pi(F) \otimes \Gamma^{(3)}(F)\Pi(G) \otimes \Gamma^{(3)}(G) \cdot \Pi(F^{-1}) \otimes \Gamma^{(3)}(F^{-1}) \\ &= \hat{P}_F\hat{P}_G\hat{P}_G^{-1}\{\widetilde{X}_k(\xi)\}\Pi(F^{-1}) \otimes \Gamma^{(3)}(F^{-1}) \\ &= \hat{P}_F\{\widetilde{X}_k(\xi)\}\Pi(F^{-1}) \otimes \Gamma^{(3)}(F^{-1}) \\ &= \{\widetilde{X}_k(\xi)\} \end{aligned} \tag{2.71′}$$

Thus

$$\begin{aligned} &\Pi(F)\Pi(G)\Pi(F^{-1}) \otimes \Gamma^{(3)}(F)\Gamma^{(3)}(G)\Gamma^{(3)}(F^{-1}) = \\ &= \Pi(G') \otimes \Gamma^{(3)}(G') \in \Gamma^{(NCf)}\{\mathscr{G}\} \end{aligned} \tag{2.72}$$

$$\Pi(F)\Pi(G)\Pi(F^{-1}) = \Pi(G') \in \Pi\{\mathscr{G}\} \tag{2.72′}$$

$$\Gamma^{(3)}(F)\Gamma^{(3)}(G)\Gamma^{(3)}(F^{-1}) = \Gamma^{(3)}(G') \in \Gamma^{(3)}\{\mathscr{G}\} \tag{2.72″}$$

The set of solutions (2.70) defines the representation $\Gamma^{(NCf)}\{\mathscr{H}\}$ of the full isometric group $\mathscr{H}(\xi)$ on the nuclear position vectors referred to the frame system

$$\Gamma^{(NCf)}\{\mathscr{H}\} := \{\Pi(H) \otimes \Gamma^{(3)}(H) \mid \forall H \in \mathscr{H}(\xi)\} \overset{is}{=} \mathscr{H}(\xi) \tag{2.73}$$

The decomposition of $\Gamma^{(NCf)}\{\mathscr{H}\}$ modulo the invariant subgroup $\Gamma^{(NCf)}\{\mathscr{G}\}$ defines a factor group isomorphic to the internal isometric group $\mathscr{F}(\xi)$

$$\frac{\Gamma^{(NCf)}\{\mathscr{H}\}}{\Gamma^{(NCf)}\{\mathscr{G}\}} = \bigcup_{k=1}^{|\mathscr{F}|} \Gamma^{(NCf)}\{\mathscr{G}\}\Gamma^{(NCf)}(F_k) \overset{is}{=} \mathscr{F}(\xi) \tag{2.74}$$

As will be shown below [Eq. (2.80)] there exists always a subgroup $\Gamma^{(NCf)}\{\mathscr{F}\}$ in $\Gamma^{(NCf)}\{\mathscr{H}\}$ isomorphic to this factor group. Therefore, Eq. (2.74) suggests the following important theorem:
The representation $\Gamma^{(NCf)}\{\mathscr{H}\}$ (abstract group $\mathscr{H}$) is a semidirect product of $\Gamma^{(NCf)}\{\mathscr{G}\}$ (abstract group $\mathscr{G}$) and $\Gamma^{(NCf)}\{\mathscr{F}\}$ (abstract group $\mathscr{F}$)

$$\begin{gathered}\Gamma^{(NCf)}\{\mathscr{G}\}\Gamma^{(NCf)}\{\mathscr{F}\} = \Gamma^{(NCf)}\{\mathscr{F}\}\Gamma^{(NCf)}\{\mathscr{G}\} = \Gamma^{(NCf)}\{\mathscr{H}\}\\ \mathscr{G}(\xi)\cdot\mathscr{F}(\xi) = \mathscr{F}(\xi)\cdot\mathscr{G}(\xi) = \mathscr{H}(\xi)\end{gathered} \tag{2.75}$$

The proof of this theorem rests on the homomorphism

$$\begin{gathered}\hat{\eta} : \Gamma^{(NCf)}\{\mathscr{H}\} \longrightarrow \Gamma^{(3)}\{\mathscr{L}\}\\ \ker\hat{\eta} := \{\Pi(H) \otimes 1^{(3)} \mid H \in \mathscr{H}(\xi)\}\end{gathered} \tag{2.76}$$

where $\Gamma^{(3)}\{\mathscr{L}\}$ (abstract group $\mathscr{L}$) denotes the set of all different rotational parts $\Gamma^{(3)}(H)$ of $\Gamma^{(NCf)}\{\mathscr{H}\}$

$$\Gamma^{(3)}\{\mathscr{L}\} := \{\Gamma^{(3)}(H) \mid \forall H \in \mathscr{H}(\xi)\} \tag{2.77}$$

The homomorphism (2.76) maps the invariant subgroup $\Gamma^{(NCf)}\{\mathscr{G}\}$ onto $\Gamma^{(3)}\{\mathscr{G}\}$, which by virtue of Eq. (2.72″) and homomorphy theorems must be a normal subgroup of $\Gamma^{(3)}\{\mathscr{L}\}$

$$\Gamma^{(3)}\{\mathscr{G}\} \overset{i}{\subset} \Gamma^{(3)}\{\mathscr{L}\} \tag{2.78}$$

Since $\Gamma^{(3)}\{\mathscr{L}\}$ is a point symmetry group (subgroup of $O(3)$), it has the property to be generated as a semidirect product of two subgroups

$$\begin{gathered}\Gamma^{(3)}\{\mathscr{L}\} = \Gamma^{(3)}\{\mathscr{G}\}\Gamma^{(3)}\{\mathscr{K}\}\\ \frac{\Gamma^{(3)}\{\mathscr{L}\}}{\Gamma^{(3)}\{\mathscr{G}\}} \overset{is}{=} \Gamma^{(3)}\{\mathscr{K}\} \subset \Gamma^{(3)}\{\mathscr{L}\}\end{gathered} \tag{2.79}$$

i.e. $\Gamma^{(3)}\{\mathscr{K}\}$ is an endomorphism of $\Gamma^{(3)}\{\mathscr{L}\}$. The inverse homomorphism (2.76) defines the representation $\Gamma^{(NCf)}\{\mathscr{F}\} \overset{is}{=} \mathscr{F}(\xi)$

$$\hat{\eta}^{-1}(\Gamma^{(3)}\{\mathscr{K}\}) = \Gamma^{(\mathrm{NCf})}\{\mathscr{F}\} \subset \Gamma^{(\mathrm{NCf})}\{\mathscr{H}\} \tag{2.80}$$

On the other hand, by the first isomorphy theorem[22)]

$$\hat{\eta}^{-1}(\Gamma^{(3)}\{\mathscr{G}\}) = \Gamma^{(\mathrm{NCf})}\{\mathscr{G}\} \cdot \ker \hat{\eta} \tag{2.81}$$

Eqs. (2.76)–(2.81) imply that

$$\begin{gathered}\Gamma^{(\mathrm{NCf})}\{\mathscr{G}\} \cdot \Gamma^{(\mathrm{NCf})}\{\mathscr{F}\} = \Gamma^{(\mathrm{NCf})}\{\mathscr{H}\} \\ \frac{\Gamma^{(\mathrm{NCf})}\{\mathscr{H}\}}{\Gamma^{(\mathrm{NCf})}\{\mathscr{G}\}} \overset{\mathrm{is}}{=} \Gamma^{(\mathrm{NCf})}\{\mathscr{F}\} \text{ Q.E.D.}\end{gathered} \tag{2.82}$$

From Eq. (2.82) the semidirect product structure of any other of the representations of $\mathscr{H}$ listed in Fig. 1 may be derived by the first isomorphy theorem[22)]. For example for the representation

$$\Gamma\{\mathscr{H}\} := \left\{ \begin{bmatrix} \epsilon' \\ \xi' \\ 1 \end{bmatrix} = \begin{bmatrix} B(H) & 0 & b(H) \\ . & A(H) & a(H) \\ . & . & 1 \end{bmatrix} \begin{bmatrix} \epsilon \\ \xi \\ 1 \end{bmatrix} \mid \forall H \in \mathscr{H}(\xi) \right\} \tag{2.83}$$

i.e. the transformation group of the eulerian angles and the internal coordinates we have

$$\Gamma\{\mathscr{G}\} \cdot \Gamma\{\mathscr{F}\} = \Gamma\{\mathscr{F}\} \cdot \Gamma\{\mathscr{G}\} = \Gamma\{\mathscr{H}\} \tag{2.84}$$

The group $\Gamma\{\mathscr{H}\}$ will be found important for the symmetry of the hamiltonian of the rotation-internal nuclear motion problem associated with SRMs (Sect. 3.2). In particular its homomorphism to $\Gamma^{(3)}\{\mathscr{L}\}$

$$\Gamma\{\mathscr{H}\} \overset{\mathrm{ho}}{=} \mathscr{B}\{\mathscr{H}\} \overset{\mathrm{is}}{=} \Delta^{(3)}\{\mathscr{L}\} \overset{\mathrm{ho}}{=} \Gamma^{(3)}\{\mathscr{L}\} \tag{2.85}$$

will be relevant for a general formulation of Wigner-Eckart type theorems for irreducible tensor operators connected with SRMs, cf. Sect. 3.3. For $\Gamma^{(3)}\{\mathscr{L}\}$ we may distinguish again SRMs of case a, b1 and b2 in strict analogy to $\Gamma^{(3)}\{\mathscr{K}\}$ Eqs. (2.18)–(2.22). Therefore, for case b SRMs

$$\Gamma\{\mathscr{K}\} \overset{\mathrm{ho}}{=} \mathscr{V}_2 \tag{2.86}$$

2.3 Relation Between the Isometric Group and the Permutation-Inversion Group (Longuet-Higgins Group) of Nonrigid Molecules

In this section a relation of the isometric group approach to the permutation-inversion group of nonrigid molecules, introduced by Longuet-Higgins[7)], will be established. Such an interrelation is obtained in a natural way, if the isometric transformations are applied to the substrate consisting of the coordinate vectors $X_k^l(\epsilon, \xi)$ referred to the LS $\tilde{e}^l$. According to Eq. (2.1)

$$X_k^l(\epsilon, \xi) = \tilde{R}(\epsilon) X_k(\xi) \tag{2.87}$$

2.3.1 SRMs with $\mathscr{G}(\xi) = C_1$

Using the general transformation formula for rotation group coefficients derived in Appendix 2

$$\hat{P}_F R(\epsilon) = \tilde{R}(F) R(\epsilon) \tag{2.88}$$

we find for any $F \in \mathscr{F}$

$$\begin{aligned}
\hat{P}_F \{\tilde{X}_k^l(\epsilon, \xi)\} &= \hat{P}_F \{\tilde{X}_k(\xi)\}\ 1^{(K)} \otimes R(\epsilon) \\
&= \{\tilde{X}_k(\xi)\}\ (\Pi(F) \otimes \Gamma^{(3)}(F))(1^{(K)} \otimes \tilde{R}(F) R(\epsilon)) \\
&= \{\tilde{X}_k^l(\epsilon, \xi)\}\ (1^{(K)} \otimes \tilde{R}(\epsilon))(\Pi(F) \otimes \Gamma^{(3)}(F))\ (1^{(K)} \otimes \tilde{R}(F) R(\epsilon)) \\
&= \{\tilde{X}_k^l(\epsilon, \xi)\} \Pi(F) \otimes \tilde{R}(\epsilon) \Gamma^{(3)}(F) \tilde{R}(F) R(\epsilon) \\
&= \{\tilde{X}_k^l(\epsilon, \xi)\} \Pi(F) \otimes 1^{(3)} | \Gamma^{(3)}(F) | = \{\tilde{X}_k^l(\epsilon, \xi)\} \Gamma^{(NCl)}(F)
\end{aligned} \tag{2.89}$$

since $\Gamma^{(3)}(F) \tilde{R}(F) = |\Gamma^{(3)}(F)|\ 1^{(3)}$

The index l indicates that this representation refers to the position vectors expressed in the laboratory coordinate system. The last equation may be commented upon as follows:

(i) if $\Gamma^{(3)}(F)$ is properly orthogonal, the isometric transformation $F \in \mathscr{F}(\xi)$ induces on the substrate $\{\tilde{X}_k^l\}$ merely a permutation of the coordinate vectors of a set of equivalent nuclei.

(ii) if $\Gamma^{(3)}(F)$ is improperly orthogonal, i.e. if $|\Gamma^{(3)}(F)| = -1$, $F \in \mathscr{F}$ induces a permutation and an inversion of the coordinate vectors of a set of equivalent nuclei. Hence, the representation

$$\Gamma^{(NCl)}\{\mathscr{F}\} := \{\Pi(F) \otimes |\Gamma^{(3)}(F)|\ 1^{(3)} \mid \forall F \in \mathscr{F}(\xi)\} \tag{2.90}$$

is an analogue of the permutation-inversion group $\mathscr{L}\mathscr{H}$ (Longuet-Higgins group). More explicitely we have the mapping expressed in Table 4.

2.3.2 SRMs with Proper Covering Group $\mathscr{G}(\xi)$

Application of the operators $\hat{P}_G$ to $\{\tilde{X}_k^l(\epsilon, \xi)\}$ gives

$$\hat{P}_G \{\tilde{X}_k^l(\epsilon, \xi)\} = \{\tilde{X}_k(\xi)\} \Gamma^{(NCf)}(G)\ 1^{(K)} \otimes \tilde{R}(G) R(\epsilon) \tag{2.91}$$

Use of the Eq. (2.49″) for $\Gamma^{(NCf)}(G)$ leads, in strict analogy to Eq. (2.89) to the set

Table 4. Relation between the isometric group and the permutation-inversion group

$\lvert\Gamma^{(3)}(F)\rvert$	$\Gamma^{(NCI)}(F)$	$\mathcal{LH}$
1	$1^{(K)} \otimes 1^{(3)}$	E
1	$\Pi(F) \otimes 1^{(3)}$	P
−1	$1^{(K)} \otimes (-1^{(3)})$	E*
−1	$\Pi(F) \otimes (-1^{(3)})$	P*

$$\Gamma^{(NCI)}\{\mathcal{G}\} := \{\Pi(G) \otimes \lvert\Gamma^{(3)}(G)\rvert\, 1^{(3)} \mid \forall G \in \mathcal{G}(\xi)\} \tag{2.92}$$

It is of the same form as $\Gamma^{(NCI)}\{\mathcal{F}\}$ and forms the analogue of the Longuet-Higgins permutation-inversion operations associated with covering symmetry operations of a SRM.

The semidirect product of the two groups (2.90) and (2.92) may be considered as the analogue of the permutation-inversion group $\mathcal{LH}$ for molecules with non-trivial covering group $\mathcal{G}(\xi)$

$$\begin{aligned}&\Gamma^{(NCI)}\{\mathcal{G}\}\,\Gamma^{(NCI)}\{\mathcal{F}\} = \Gamma^{(NCI)}\{\mathcal{H}\} \overset{\text{is}}{=} \mathcal{LH}\\&\Gamma^{(NCI)}\{\mathcal{H}\} := \{\Pi(H) \otimes \lvert\Gamma^{(3)}(H)\rvert\, 1^{(3)} \mid \forall H \in \mathcal{H}(\xi)\}\end{aligned} \tag{2.93}$$

2.3.3 Primitive Period Isometric Transformations and the Longuet-Higgins Group

It is of interest to point out the role of primitive period isometric transformations (cf. Sect. 2.2.2.) in both the isometric and the Longuet-Higgins group. According to Eq. (2.41) this type of transformations is represented on the basis $\{\underset{\sim}{X}_k(\xi)\}$ by

$$\Gamma^{(NCf)}(F_p) = 1^{(K)} \otimes \Gamma^{(3)}(F_p) \tag{2.94}$$

Considering now the action of primitive period isometric operators on the basis $\{\widetilde{X}^l_k(\epsilon, \xi)\}$, we will distinguish between the following cases

(i) if $\Gamma^{(3)}(F_p) \in SO(3)$, then it follows from Eqs. (2.89) and (2.94) that

$$\begin{aligned}\hat{P}_{F_p}\{\widetilde{X}^l_k(\epsilon, \xi)\} &= \{\widetilde{X}^l_k(\epsilon, \xi)\}\, 1^{(K)} \otimes \lvert\Gamma^{(3)}(F_p)\rvert\, 1^{(3)}\\&= \{\widetilde{X}^l_k(\epsilon, \xi)\}\, 1^{(K)} \otimes 1^{(3)}\end{aligned} \tag{2.95}$$

This shows that the representation of the group $\mathcal{F}(\xi)$ on the nuclear position vectors referred to the laboratory fixed coordinate system is not a faithful representation of the isometric group $\overline{\mathcal{F}}(\xi) \overset{\text{is}}{=} \Gamma^{(NCf)}\{\overline{\mathcal{F}}\}$

$$\overline{\mathcal{F}}(\xi) \overset{\text{ho}}{=} \Gamma^{(NCI)}\{\overline{\mathcal{F}}\} \overset{\text{is}}{=} \mathcal{F}(\xi) \tag{2.96}$$

Therefore, SRMs with $\mathcal{G}(\xi) = \underline{C_1}$, for which the permutation-inversion group is always $\underline{\text{is}}$omorphic to $\Gamma^{(NCI)}\{\overline{\mathcal{F}}\}$ the relation between $\mathcal{LH}$ and the isometric group $\overline{\mathcal{F}}(\xi)$ is only a homomorphism

$$\mathscr{L}\mathscr{H} \overset{\text{is}}{=} \Gamma^{(\text{NCl})}\{\overline{\mathscr{F}}\} \overset{\text{ho}}{=} \overline{\mathscr{F}}(\xi) \tag{2.97}$$

The kernel of the homomorphism (2.97) is the subgroup of $\overline{\mathscr{F}}(\xi)$ consisting of the set of all primitive period transformations. This explains why in some applications to the nuclear motion problem of nonrigid molecules a "double group" of the Longuet-Higgins group had to be used[23, 24].

(ii) If $\Gamma^{(3)}(F_p)$ is an element of the coset $Z \cdot SO(3)$ of the decomposition $O(3) = SO(3) \cup Z \cdot SO(3)$ one has

$$\Gamma^{(\text{NCl})}(F_p) = 1^{(K)} \otimes (-1^{(3)}) \tag{2.98}$$

Hence,

$$\begin{aligned} &\mathscr{L}\mathscr{H} \overset{\text{is}}{=} \Gamma^{(\text{NCl})}\{\overline{\mathscr{F}}\} \overset{\text{ho}}{=} \overline{\mathscr{F}}(\xi) \\ &\text{if } 1^{(K)} \otimes (-1^{(3)}) \in \Gamma^{(\text{NCl})}\{\overline{\mathscr{F}}\} \end{aligned} \tag{2.99}$$

and

$$\begin{aligned} &\mathscr{L}\mathscr{H} \overset{\text{is}}{=} \Gamma^{(\text{NCl})}\{\overline{\mathscr{F}}\} \overset{\text{is}}{=} \overline{\mathscr{F}}(\xi) \\ &\text{if } 1^{(K)} \otimes (-1^{(3)}) \notin \Gamma^{(\text{NCl})}\{\overline{\mathscr{F}}\} \end{aligned} \tag{2.99'}$$

although no example of type (2.99′) is known.

Analogously for SRMs with proper covering group $\mathscr{G}(\xi)$ and primitive period transformations, the full isometric group $\overline{\mathscr{H}}(\xi) = \overline{\mathscr{F}}(\xi) \cdot \mathscr{G}(\xi)$ is homomorphic to the permutation-inversion group

$$\mathscr{L}\mathscr{H} \overset{\text{is}}{=} \Gamma^{(\text{NCl})}\{\overline{\mathscr{H}}\} \overset{\text{ho}}{=} \overline{\mathscr{H}}(\xi) \tag{2.100}$$

Again it has been shown that in these cases introduction of a "double group" corresponds to extension of the isometric group $\mathscr{H}(\xi)$ to $\overline{\mathscr{H}}(\xi)$[25], the latter being a symmetry of the rotation-internal nuclear motion hamiltonian.

2.4 Examples for Isometric Groups of SRMs

A considerable number of groups of nonrigid molecules has been discussed in the literature[26–28]. An attempt for a systematic classification of isometric groups has been reported for the first time by Frei et al.[15]. In order to illustrate the construction principles given in Sect. 2.2. a few examples will now be discussed. The examples are listed in Table 5 and chosen such that specific aspects both of the construction process and the group structure may be emphasized. In the table a symbol for the SRM defining frame, tops, etc. and the respective local symmetries, the number of finite internal coordinates, the covering symmetry group $\mathscr{G}(\xi)$ and one representative molecule are given.

Table 5. Examples for isometric groups of SRMs

SRM	f	$\mathscr{G}(\xi)$	Typical molecule
$D_{\infty h}F(C_1TR)(C_1TS)$	1	C_1	(1-R,2-S)–CHFCl–CHFCl
$C_2(\tau)F(C_sT)_2$	3	C_1	$CH_2OH–CH_2OH$
$D_{\infty h}F(C_1TR)(C_1TS)$	3	C_1	$CHFCl–(C_6H_4)_2–CHFCl$
$D_{\infty h}F(C_1TR)_2$			
$D_{\infty h}F(C_{2v}T)_2$	1	D_2	$(C_6H_5)_2$

2.4.1 $D_{\infty h}F(C_1TR)(C_1TS)$ System

As a first example we consider a very simple SRM without covering symmetry ($\mathscr{G}(\xi) = C_1$) which allows to show in detail all steps of the construction of the internal isometric group $\mathscr{F}$ and to demonstrate the effect of primitive period isometric transformations. Figure 2 shows Newman projections of (1–R, 2–S)–CHFCl–CHFCl, a molecule with a rigid C–C frame of symmetry $D_{\infty h}$ to which two equivalent CHFCl tops of local symmetry C_1 with opposite configuration are attached. The molecule

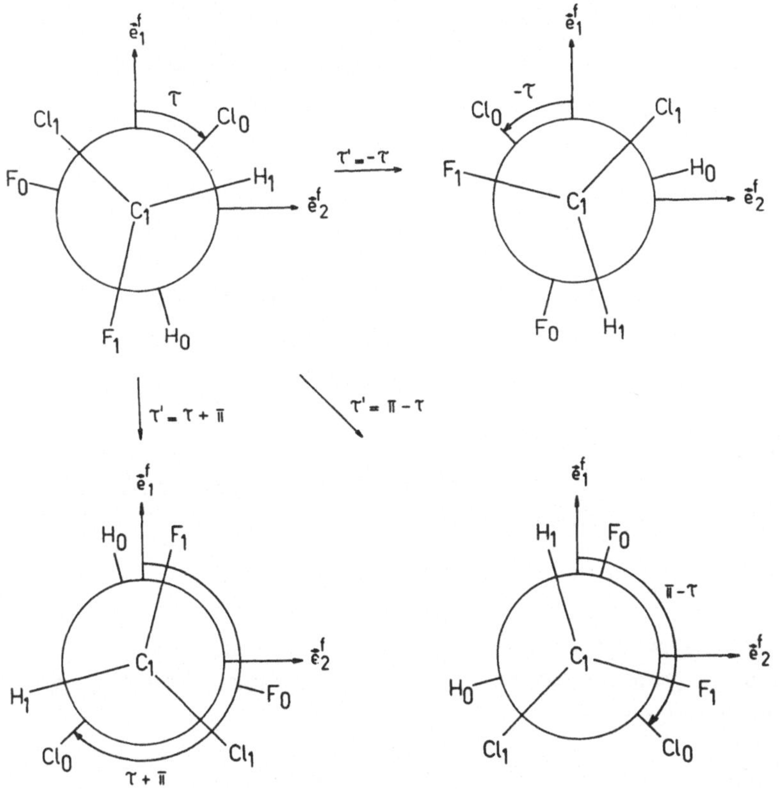

Fig. 2. Newman projection of $D_{\infty h}F(C_1TR)(C_1TS)$ system
Key: All internal isometric transformations, including primitive period transformation are shown

Table 6. Coordinate vectors of equivalent nuclei of $D_{\infty h}F(C_1TR)(C_1TS)$ system

Set	Vector[a]	X
Frame $\lambda = 0,1$	$X_{f\lambda}$	$\frac{1}{2} r \begin{bmatrix}1 & & \\ & -1 & \\ & & -1\end{bmatrix}^{\lambda} \cdot \begin{bmatrix}0\\0\\1\end{bmatrix}$
Top λ $\lambda = 0,1$	$X_{t\lambda}$	$\begin{bmatrix}1 & & \\ & -1 & \\ & & -1\end{bmatrix}^{\lambda} \left\{ \frac{1}{2} r \begin{bmatrix}0\\0\\1\end{bmatrix} + \begin{bmatrix}c\tau & -s\tau & 0\\ s\tau & c\tau & 0\\ 0 & 0 & 1\end{bmatrix} \begin{bmatrix}1 & & \\ & -1 & \\ & & 1\end{bmatrix}^{\lambda} X_0^{t\,b} \right\}$

[a] Coordinates refer to the frame coordinate system $\tilde{e}^f$ indicated in Fig. 2; the structural parameter r denotes the C–C bond length.

[b] Coordinate vector of representative nucleus w.r.t. local top coordinate system $\tilde{e}^t$ whose origin lies in the nucleus C_0 and whose axis e_3^t coincides with e_3^f.

fixed coordinate system and the internal coordinate τ are defined in the figure. The axis e_3^f coincides with the C–C bond, whereas the axis e_1^f bisects the dihedral angle 2τ. The origin of the frame coordinate system lies in the center of the CC bond. Table 6 shows the coordinate vectors of equivalent frame and top nuclei referred to the frame coordinate system. From these coordinate vectors, the following formula for the distances between e.g. Cl and H nuclei of opposite tops may be obtained (c: cos, s: sin)

$$\begin{aligned} d^2_{Cl\lambda,H\lambda+1}(\tau) &= |X_{Cl\lambda}(\tau) - X_{H\lambda+1}(\tau)|^2 \\ &= r^2 + 2\,rX^t_{Cl0\,3} + 2\,rX^t_{H0\,3} + \tilde{X}^t_{Cl0}X^t_{H0} + \tilde{X}^t_{H0}X^t_{H0} \\ &\quad -2\left(X^t_{Cl0\,1} \quad 0 \quad X^t_{Cl0\,3}\right) \begin{bmatrix} c2\tau & -s2\tau & 0\\ -s2\tau & -c2\tau & 0\\ 0 & 0 & -1\end{bmatrix} \begin{bmatrix} X^t_{H0\,1}\\ (-1)^{\lambda+1}\,X^t_{H0\,2}\\ X^t_{H0\,3}\end{bmatrix} \\ &\qquad \lambda = 0,1 \end{aligned}$$

This formula shows that the primitive period p of τ is equal to π. If we take the domain $-\pi/2 < \tau \leqslant +\pi/2$, the only nontrivial internal isometric transformation of this SRM is $F_2 : \tau' = -\tau$ and therefore, since $F_2^2 = E$

$$\mathscr{A}\{\mathscr{F}\} := \{\mathscr{A}(E), \mathscr{A}(F_2)\} \stackrel{\text{is}}{=} \Gamma^{(\mathscr{N}\mathscr{C})}\{\mathscr{F}\} \stackrel{\text{is}}{=} \mathscr{F}(\tau) \stackrel{\text{is}}{=} \mathscr{C}_2$$

cf. Table 7. The internal isometric transformation F_2 is visualized in Fig. 2.

Application of the primitive period $\tau' = \tau + \pi$ to the coordinate vectors of a set of equivalent top nuclei gives

$$\hat{P}_{F_p}\{\tilde{X}_{t\lambda}(\tau)\} = \{\tilde{X}_{t\lambda}(\tau - \pi)\} = \{\tilde{X}_{t\lambda}(\tau)\}\, 1^{(2)} \otimes \begin{pmatrix} -1 & & \\ & -1 & \\ & & 1\end{pmatrix} \neq \{\tilde{X}_{t\lambda}(\tau)\}$$

$$\Gamma^{(\mathscr{N}\mathscr{C})}(F_p) = 1^{(2)}$$

Table 7. Isometric group of the $D_{\infty h}F(C_1TR)(C_1TS)$ system $\mathcal{G}(\tau) = C_1$, $\overline{\mathcal{F}}(\tau) \overset{\text{is}}{=} \mathcal{V}_4$, $\mathcal{F}(\tau) \overset{\text{is}}{=} \mathcal{V}_2$

Operator $F \in \overline{\mathcal{F}}$	$\mathcal{A}\{\overline{\mathcal{F}}\}^a$	$\Gamma^{(\mathcal{NC})}\{\mathcal{F}\}^b$	$\Gamma^{(NCf)}\{\overline{\mathcal{F}}\}^c$	$\Gamma\{\overline{\mathcal{F}}\}^d$
E	$\begin{bmatrix} 1 & . \\ . & 1 \end{bmatrix}$	$\begin{bmatrix} 1 & . \\ . & 1 \end{bmatrix}$	$\begin{bmatrix} 1 & . \\ . & 1 \end{bmatrix} \otimes \begin{bmatrix} 1 & & \\ & 1 & \\ & & 1 \end{bmatrix}$	$\begin{bmatrix} 1 & . & . & . & . \\ & 1 & . & . & . \\ & & 1 & . & . \\ & & & 1 & . \\ & & & & 1 \end{bmatrix}$
F_2	$\begin{bmatrix} -1 & . \\ . & 1 \end{bmatrix}$	$\begin{bmatrix} . & 1 \\ 1 & . \end{bmatrix}$	$\begin{bmatrix} . & 1 \\ 1 & . \end{bmatrix} \otimes \begin{bmatrix} 1 & & \\ & 1 & \\ & & -1 \end{bmatrix}$	$\begin{bmatrix} 1 & . & . & . & . \\ & 1 & . & . & . \\ & & 1 & . & \pi \\ & & & -1 & . \\ & & & & 1 \end{bmatrix}$
$F_3 = F_p$	$\begin{bmatrix} 1 & \pi \\ . & 1 \end{bmatrix}$	$\begin{bmatrix} 1 & . \\ . & 1 \end{bmatrix}$	$\begin{bmatrix} 1 & . \\ . & 1 \end{bmatrix} \otimes \begin{bmatrix} -1 & & \\ & -1 & \\ & & 1 \end{bmatrix}$	$\begin{bmatrix} 1 & . & . & . & . \\ & 1 & . & . & . \\ & & 1 & . & \pi \\ & & & 1 & \pi \\ & & & & 1 \end{bmatrix}$
$F_4 = F_2 \cdot F_p$	$\begin{bmatrix} -1 & \pi \\ . & 1 \end{bmatrix}$	$\begin{bmatrix} . & 1 \\ 1 & . \end{bmatrix}$	$\begin{bmatrix} . & 1 \\ 1 & . \end{bmatrix} \otimes \begin{bmatrix} 1 & & \\ & -1 & \\ & & -1 \end{bmatrix}$	$\begin{bmatrix} 1 & . & . & . & . \\ & 1 & . & . & . \\ & & 1 & . & . \\ & & & -1 & \pi \\ & & & & 1 \end{bmatrix}$

[a] Representation of $\overline{\mathcal{F}}$ by substitutions of the internal coordinate:

$$\begin{bmatrix} \tau' \\ 1 \end{bmatrix} = \begin{bmatrix} A(F) & a(F) \\ 0 & 1 \end{bmatrix} \cdot \begin{bmatrix} \tau \\ 1 \end{bmatrix}, \quad -\pi < \tau \leqslant +\pi, \ \tau \bmod 2\,p.$$

[b] Representation by the set of distances $d_{Cl_0,\,H_1}(\tau)$, $d_{Cl_1,\,H_0}(\tau)$ originating from the two equivalent tops (see Fig. 2).

[c] Representation generated by the vectors $\tilde{X}_{t0}(\tau)$, $\tilde{X}_{t1}(\tau)$ of a set of equivalent nuclei originating from the two equivalent tops (see Fig. 2).

[d] Representation of $\overline{\mathcal{F}}$ by substitutions of the eulerian angles and internal coordinates:

$$\begin{bmatrix} \alpha' \\ \beta' \\ \gamma' \\ \tau' \\ 1 \end{bmatrix} = \begin{bmatrix} B(F) & 0 & b(F) \\ . & A(F) & a(F) \\ . & . & 1 \end{bmatrix} \begin{bmatrix} \alpha \\ \beta \\ \gamma \\ \tau \\ 1 \end{bmatrix}$$

cf. Fig. 2. Therefore, to get the symmetry group $\overline{\mathcal{F}}$ of the rotation-internal motion problem, $\mathcal{F}(\tau)$ has to be extended by F_p ($\mathcal{V}_4$ denotes the four group)

$$\mathcal{A}\{\overline{\mathcal{F}}\} \overset{\text{is}}{=} \overline{\mathcal{F}}(\tau) := \{E, F_2, F_p, F_2 \cdot F_p\} \overset{\text{is}}{=} \mathcal{V}_4$$
$$\overline{\mathcal{F}} \overset{\text{end}}{=} \mathcal{F}$$

The representation of $\overline{\mathcal{F}}$ on the nuclear coordinate vectors is obtained by solving Eq. (2.12) for all $F \in \overline{\mathcal{F}}$

$$\hat{P}_F\{\tilde{X}_{t0}(\tau)\tilde{X}_{t1}(\tau)\} = \{\tilde{X}_{t0}(F^{-1}(\tau))\tilde{X}_{t1}(F^{-1}(\tau))\} = \\ = \{\tilde{X}_{t0}(\tau)\tilde{X}_{t1}(\tau)\}\Pi(F)\otimes\Gamma^{(3)}(F)$$

or (for one position vector)

$$\hat{P}_F\tilde{X}_{t\lambda}(\tau) = \tilde{X}_{t\lambda}(F^{-1}(\tau)) = \tilde{X}_{t\lambda'}(\tau)\Gamma^{(3)}(F) \tag{2.12'}$$

e.g. for $F_2 : \tilde{X}_{t\lambda}(-\tau) = \tilde{X}_{t\lambda'}(\tau)\Gamma^{(3)}(F_2)$.

Explicitely

$$\left\{\frac{1}{2}r(0\,0\,1) + \tilde{X}_0^t\begin{bmatrix}1 & & \\ & (-1)^\lambda & \\ & & 1\end{bmatrix}\begin{bmatrix}c\tau & -s\tau & 0\\ s\tau & c\tau & 0\\ 0 & 0 & 1\end{bmatrix}\right\}\begin{bmatrix}1 & & \\ & -1 & \\ & & -1\end{bmatrix}^\lambda$$

$$= \left\{\frac{1}{2}r(0\,0\,1) + \tilde{X}_0^t\begin{bmatrix}1 & & \\ & (-1)^{\lambda'} & \\ & & 1\end{bmatrix}\begin{bmatrix}c\tau & s\tau & 0\\ -s\tau & c\tau & 0\\ 0 & 0 & 1\end{bmatrix}\right\}\begin{bmatrix}1 & & \\ & -1 & \\ & & -1\end{bmatrix}^{\lambda'}\cdot\Gamma^{(3)}(F_2)$$

X_0^t is an arbitrary vector, therefore, solution of this equation gives the 3 by 3 matrix

$$\Gamma^{(3)}(F_2) =$$

$$\begin{bmatrix} c^2\tau + (-1)^{\lambda+\lambda'+1}\cdot s^2\tau & (-1)^{\lambda'}\cdot s\tau\cdot c\tau + (-1)^\lambda\cdot s\tau\, c\tau & 0\\ (-1)^{\lambda+1}\cdot s\tau\cdot c\tau + (-1)^{\lambda'+1}\cdot s\tau\, c\tau & (-1)^{\lambda+\lambda'+1}\cdot s^2\tau + c^2\tau & 0\\ 0 & 0 & (-1)^{\lambda+\lambda'}\end{bmatrix}$$

Since $\Gamma^{(3)}(F_2)$ must be independent of τ and λ and λ' (all nuclear coordinate vectors experience the same rotation $\Gamma^{(3)}(F)$), the unique solution is $\lambda' = \lambda + 1$. Therefore,

$$\Pi(F_2) = (\delta_{\lambda',\lambda+1}) = \begin{pmatrix}\cdot & 1\\ 1 & \cdot\end{pmatrix},\quad \Gamma^{(3)}(F_2) = \begin{bmatrix}1 & & \\ & 1 & \\ & & -1\end{bmatrix}$$

A more direct method to get the representation $\Gamma^{(NCf)}(F)$ uses the fact that the rotative parts $\Gamma^{(3)}(F)$ are orthogonal transformations which map the reference NC on all isometric NCs. As may be seen from Fig. 2, these are

$$F_2 : S_{12}^f,\ F_3 : C_2(e_3^f),\ F_4 : Z$$

The permutation matrices $\Pi(F)$ are then obtained by solving the Eqs. (2.12').

For very simple SRMs $\Gamma^{(NCf)}(F)$ may be constructed by means of a drawing or a molecular model, because to any linear operator $\hat{P}_F$ there is associated a mapping (in the fixed frame system) which can be determined by geometrical reasoning.

The representation $\Gamma^{(NCf)}\{\overline{\mathscr{F}}\}$ together with representations of $\overline{\mathscr{F}}$ on other substrates is collected in Table 7. From $\Gamma^{(NCf)}\{\overline{\mathscr{F}}\}$ one gets

$$\Pi\{\overline{\mathcal{F}}\} := \left\{\begin{pmatrix}1 & \cdot\\ \cdot & 1\end{pmatrix}, \begin{pmatrix}\cdot & 1\\ 1 & \cdot\end{pmatrix}\right\} \stackrel{\text{is}}{=} \mathcal{V}_2$$

$$\Gamma^{(3)}\{\overline{\mathcal{K}}\} := \left\{\begin{pmatrix}1 & & \\ & 1 & \\ & & 1\end{pmatrix}, \begin{pmatrix}1 & & \\ & 1 & \\ & & -1\end{pmatrix}, \begin{pmatrix}-1 & & \\ & -1 & \\ & & 1\end{pmatrix}, \begin{pmatrix}-1 & & \\ & -1 & \\ & & -1\end{pmatrix}\right\} = C_{2h}(e_3^f) \stackrel{\text{is}}{=} \mathcal{V}_4$$

Z explicitely occurs in $\Gamma^{(3)}\{\overline{\mathcal{K}}\}$, therefore the set of matrices $R(F) = |\Gamma^{(3)}(F)| \cdot \Gamma^{(3)}(F)$ form the group

$$\Delta^{(3)}\{\overline{\mathcal{K}}\} \equiv \Gamma^{(3)}\{\overline{\mathcal{K}}^{+}\} := \left\{\begin{pmatrix}1 & & \\ & 1 & \\ & & 1\end{pmatrix}, \begin{pmatrix}-1 & & \\ & -1 & \\ & & 1\end{pmatrix}\right\} = C_2(e_3^f)$$

(*case b1*, cf. Sect. 2.2.1.).

By solving the Eqs. (2.26) for all $R(F) \in \Delta^{(3)}\{\overline{\mathcal{K}}^{+}\}$[5]

$$D(\epsilon') = D(\epsilon) \cdot \widetilde{R}(F)$$

one obtains the group of the transformations of the eulerian angles

$$\mathcal{B}\{\overline{\mathcal{F}}\} := \left\{\begin{bmatrix}1 & \cdot & \cdot & \cdot\\ & 1 & \cdot & \cdot\\ & & 1 & \cdot\\ & & & 1\end{bmatrix}, \begin{bmatrix}1 & \cdot & \cdot & \cdot\\ & 1 & \cdot & \cdot\\ & & 1 & \pi\\ & & & 1\end{bmatrix}\right\} \stackrel{\text{is}}{=} \mathcal{V}_2$$

and the substitution group

$$\Gamma\{\overline{\mathcal{F}}\} := \{\mathcal{B}(F) \oplus \mathcal{A}(F) \mid \forall F \in \overline{\mathcal{F}}\} \stackrel{\text{is}}{=} \mathcal{V}_4$$

cf. Table 7.

It should be remarked that the subgroup $\{E, F_4\}$ and its representations form the internal isometric group $\mathcal{F}$ for the choice $[0, \pi]$ of the domain of τ. The fixed points of the isometric transformations will be discussed in Sect. 3.4.2.

The permutation-inversion group of this SRM is only homomorphic to $\overline{\mathcal{F}}(\tau)$, since $\Gamma^{(NCl)}(F_p) = 1^{(2)} \otimes 1^{(3)}$

$$\mathcal{L}\mathcal{H} \stackrel{\text{is}}{=} \Gamma^{(NCl)}\{\overline{\mathcal{F}}\} := \{1^{(2)} \otimes 1^{(3)}, 1^{(2)} \otimes (-1^{(3)})\} \stackrel{\text{is}}{=} \mathcal{V}_2$$

$$\mathcal{L}\mathcal{H} \stackrel{\text{ho}}{=} \overline{\mathcal{F}}(\tau)$$

2.4.2 $C_2(\tau)F(C_sT)_2$ System

As a second example without covering symmetry ($\mathcal{G}(\xi) = C_1$) but several finite internal coordinates the isometric group of a semirigid model describing ethylene

---

5 The matrix $D(\epsilon)$ is explicitely given in Appendix 1.

Table 8. Coordinate vectors of equivalent nuclei of the $C_2(\tau)F(C_sT)_2$ system

Set	Vector[a]	X
Frame $\nu, \mu = 0,1$	$X_{f\nu\mu}$	$\begin{bmatrix}1&&\\&-1&\\&&-1\end{bmatrix}^{\nu}\left\{\frac{1}{2}r_1\begin{bmatrix}0\\0\\1\end{bmatrix}+\begin{bmatrix}c\tau&-s\tau&0\\s\tau&c\tau&0\\0&0&1\end{bmatrix}\cdot\begin{bmatrix}1&&\\&(-1)^{\mu}&\\&&1\end{bmatrix}X^{f}_{f00}{}^{b}\right\}$
top λ $\lambda, \kappa = 0,1$	$X_{t\lambda\kappa}$	$\begin{bmatrix}1&&\\&-1&\\&&-1\end{bmatrix}^{\lambda}\left\{\frac{1}{2}r_1\begin{bmatrix}0\\0\\1\end{bmatrix}+\begin{bmatrix}c\tau&-s\tau&0\\s\tau&c\tau&0\\0&0&1\end{bmatrix}\begin{bmatrix}c\alpha&0&s\alpha\\0&1&0\\-s\alpha&0&c\alpha\end{bmatrix}\cdot\right.$ $\left.\cdot\left[r_2\begin{bmatrix}0\\0\\1\end{bmatrix}+\begin{bmatrix}cv_\lambda&-sv_\lambda&0\\sv_\lambda&cv_\lambda&0\\0&0&1\end{bmatrix}\cdot\begin{bmatrix}1&&\\&(-1)^{\kappa}&\\&&1\end{bmatrix}X^{t}_{t00}{}^{b}\right]\right\}$

[a] Coordinates refer to the coordinate system $\tilde{e}^f$ indicated in Fig. 3; similarly the structural parameters r_1, r_2 and α are defined in Fig. 3.

[b] Coordinate vector of representative nucleus w.r.t. local coordinate system $\tilde{e}^f$ resp. $\tilde{e}^t$.

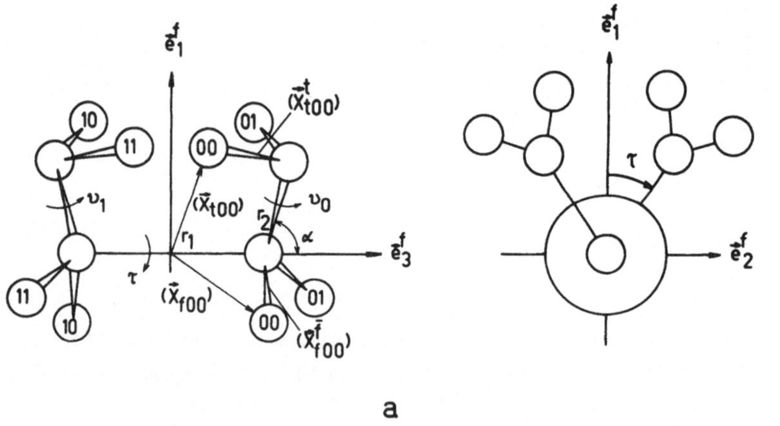

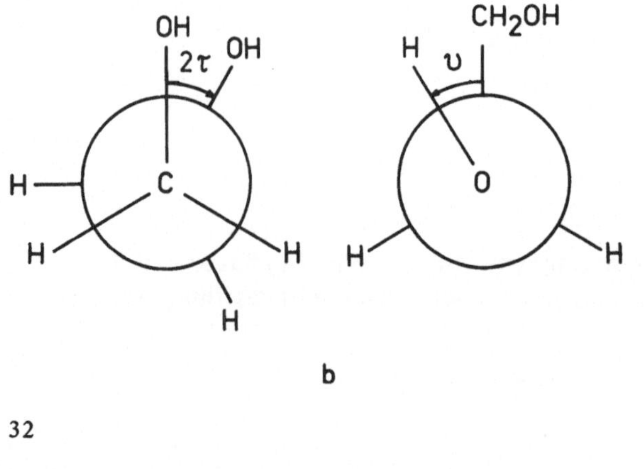

Fig. 3. Semirigid system $C_2(\tau)F(C_sT)_2$.
(a) Schematics of molecular structural parameters and coordinate systems. **(b)** Newman projections for definition of internal rotational degrees of freedom

glycol, $CH_2OH{-}CH_2OH$, will be discussed. This SRM is characterized by a semirigid $OCH_2{-}CH_2O$ frame with local symmetry C_2, and two equivalent OH tops with local symmetry C_s. The manifold of NCs of this SRM may be described by the dihedral angle 2τ of the internal rotation around the C–C bond ($\tau = 0$ for *cis* conformation) and by the two dihedral angles υ_0, υ_1 of the two OH groups ($\upsilon_0, \upsilon_1 = 0$ for *cis* conformation). The position vectors of a set of equivalent frame and top nuclei in general site[6] given in Table 8 refer to a frame fixed coordinate system whose e_3^f axis coincides with the C–C bond and whose e_1^f axis bisects the C–C bond and the dihedral angle O–C–C–O, cf. Fig. 3.

The trigonometric functions in the distance formula for the 4 distances between the 4 equivalent top nuclei

$$\{d_{t00,\,t10}(\tau, \upsilon_0, \upsilon_1)\; d_{t00,\,t11}(\tau, \upsilon_0, \upsilon_1)\; d_{t01,\,t10}(\tau, \upsilon_0, \upsilon_1)\; d_{t01,\,t11}(\tau, \upsilon_0, \upsilon_1)\}$$

$$d^2_{t0\kappa,\,t1\bar{\kappa}}(\tau, \upsilon_0, \upsilon_1) = \widetilde{(X_{t0\kappa} - X_{t1\bar{\kappa}})}(X_{t0\kappa} - X_{t1\bar{\kappa}})$$

depend on υ_0, υ_1 and 2τ. Therefore, the primitive period p of τ is equal to π. Table 9a and Fig. 4 show the four isometric transformations $\mathscr{A}(F)$

$$\begin{bmatrix} \tau' \\ \upsilon_0' \\ \upsilon_1' \\ 1 \end{bmatrix} = \mathscr{A}(F) \begin{bmatrix} \tau \\ \upsilon_0 \\ \upsilon_1 \\ 1 \end{bmatrix}$$

for the domains $-\pi/2 < \tau \leqslant +\pi/2; -\pi < \upsilon_0, \upsilon_1 \leqslant +\pi$.

They form a group isomorphic to the four group

$$\mathscr{F}(\tau, \upsilon_0, \upsilon_1) \overset{\text{is}}{=} \mathscr{A}\{\mathscr{F}\} \overset{\text{is}}{=} \Gamma^{(\mathscr{N}\mathscr{C})}\{\mathscr{F}\} \overset{\text{is}}{=} \mathscr{V}_4$$

Application of the substitution $\tau' = \tau + p$ to the position vectors of the top nuclei

$$\{\widetilde{X}_{t\lambda\kappa}(\tau, \upsilon_\lambda)\} = \{\widetilde{X}_{t00}(\tau, \upsilon_0)\widetilde{X}_{t01}(\tau, \upsilon_0)\widetilde{X}_{t10}(\tau, \upsilon_1)\widetilde{X}_{t11}(\tau, \upsilon_1)\}$$

gives

$$\hat{P}_{F_p}\{\widetilde{X}_{t\lambda\kappa}(\tau, \upsilon_\lambda)\} = \{\widetilde{X}_{t\lambda\kappa}(\tau - \pi, \upsilon_\lambda)\} = \{\widetilde{X}_{t\lambda\kappa}(\tau, \upsilon_\lambda)\}\, 1^{(4)} \otimes \begin{bmatrix} -1 & . & . \\ & -1 & . \\ & & 1 \end{bmatrix}$$

$$\Gamma^{(\mathscr{N}\mathscr{C})}(F_p) = 1^{(4)}$$

Therefore, this SRM possesses a nontrivial primitive period isometric transformation by which $\mathscr{F}$ has to be extended to get the group $\bar{\mathscr{F}}$. Since

6 Although the H nuclei of the OH groups of glycol lay in the symmetry plane of the C_s tops, it is convenient to introduce nuclei in general site in a SRM associated to a particular molecule to be sure that a faithful representation of $\mathscr{F}$ on the distances of that set is generated.

$$\mathscr{A}(\mathrm{F_p}) = \begin{bmatrix} 1 & . & . & \pi \\ & 1 & . & . \\ & & 1 & . \\ & & & 1 \end{bmatrix}$$

commutes with all $\mathscr{A}(\mathrm{F}) \in \mathscr{A}\{\mathscr{F}\}$

$$\overline{\mathscr{F}}(\tau, v_0, v_1) \overset{\text{is}}{=} \mathscr{A}\{\overline{\mathscr{F}}\} \overset{\text{is}}{=} \mathscr{A}(2, 2, 2)$$
$$\overline{\mathscr{F}} \overset{\text{end}}{=} \mathscr{F}$$

where $\mathscr{A}(2, 2, 2)$ denotes the abelian group of type 2, 2, 2 isomorphic to D_{2h}. The matrices $\mathscr{A}(\mathrm{F}) \in \mathscr{A}\{\overline{\mathscr{F}}\}$ are contained in the direct sums $\Gamma(\mathrm{F})$ listed in Table 9b. Application of the operators $\hat{\mathrm{P}}_{\mathrm{F}}$ associated with all elements $\mathrm{F} \in \mathscr{F}$ to the position vectors of the top nuclei generates the representation $\Gamma^{(\mathrm{NCf})}\{\overline{\mathscr{F}}\}$ given in Table 9b

Table 9a. Isometric group of the $C_2(\tau)F(C_sT)_2$ system
$\mathscr{G}(\tau, v_0, v_1) = C_1$, $\mathscr{F}(\tau, v_0, v_1) = \mathscr{V}_4$, $\overline{\mathscr{F}}(\tau, v_0, v_1) = \mathscr{A}(2, 2, 2)$

Operator $\mathrm{F} \in \mathscr{F}$	$\mathscr{A}\{\mathscr{F}\}^{a}$	$\Gamma^{(\mathcal{N}\mathscr{C})}\{\mathscr{F}\}^{b}$
E	$\begin{bmatrix} 1 & . & . & . \\ & 1 & . & . \\ & & 1 & . \\ & & & 1 \end{bmatrix}$	$\begin{bmatrix} 1 & . & . & . \\ & 1 & . & . \\ & & 1 & . \\ & & & 1 \end{bmatrix}$
F_2	$\begin{bmatrix} 1 & . & . & . \\ & . & 1 & . \\ & 1 & . & . \\ & & & 1 \end{bmatrix}$	$\begin{bmatrix} 1 & . & . & . \\ & & 1 & . \\ & 1 & . & . \\ & & & 1 \end{bmatrix}$
F_3	$\begin{bmatrix} -1 & . & . & . \\ & -1 & . & . \\ & & -1 & . \\ & & & 1 \end{bmatrix}$	$\begin{bmatrix} . & . & . & 1 \\ & & 1 & . \\ & 1 & . & . \\ 1 & . & . & . \end{bmatrix}$
F_4	$\begin{bmatrix} -1 & . & . & . \\ & & -1 & . \\ & -1 & . & . \\ & & & 1 \end{bmatrix}$	$\begin{bmatrix} . & . & . & 1 \\ & 1 & . & . \\ & & 1 & . \\ 1 & . & . & . \end{bmatrix}$

[a] Representation of $\mathscr{F}$ by substitutions of the internal coordinates:

$$\begin{bmatrix} \tau' \\ v_0' \\ v_1' \\ 1 \end{bmatrix} = \begin{bmatrix} \mathrm{A(F)} & \mathrm{a(F)} \\ 0 & 1 \end{bmatrix} \cdot \begin{bmatrix} \tau \\ v_0 \\ v_1 \\ 1 \end{bmatrix}, \quad \begin{matrix} -\pi/2 < \tau \leqslant +\pi/2, \tau \bmod \mathrm{p} \\ -\pi < v_0, v_1 \leqslant +\pi \end{matrix}$$

[b] Representation of $\mathscr{F}$ by permutations generated by the set of distances $d_{t00,t10}(\tau, v_0, v_1)$, $d_{t00,t11}(\tau, v_0, v_1)$, $d_{t01,t10}(\tau, v_0, v_1)$, $d_{t01,t11}(\tau, v_0, v_1)$, originating from the two equivalent tops of local symmetry C_s (see Fig. 3).

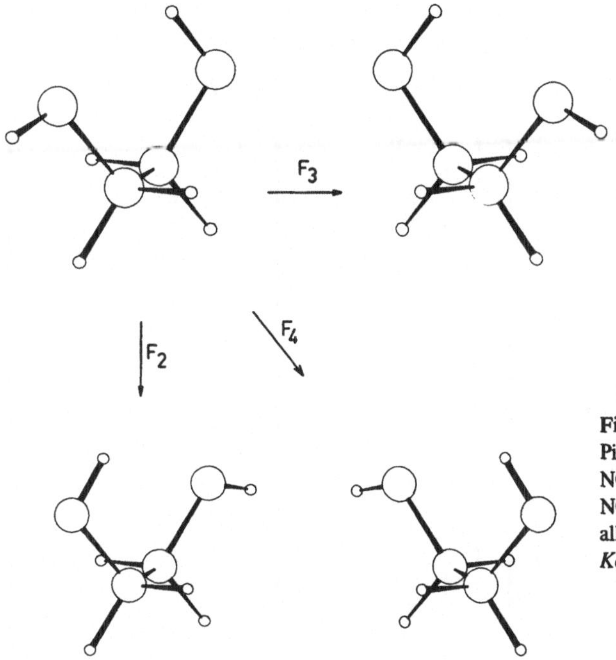

Fig. 4. $C_2(\tau)F(C_sT)_2$ system. Pictorial views showing isometric NCs generated from the reference NC on the upper lefthand side by all substitutions $F \in \mathscr{F}\ (\tau, \upsilon_0, \upsilon_1)$. *Key:* $F_2: \upsilon_0' = \upsilon_1, \upsilon_1' = \upsilon_0$; $F_3: \tau' = -\tau, \upsilon_0' = -\upsilon_0, \upsilon_1' = -\upsilon_1$; $F_4: \tau' = -\tau, \upsilon_0' = -\upsilon_1, \upsilon_1' = -\upsilon_0$

Table 9b. Isometric group of $C_2(\tau)F(C_sT)_2$ system $\mathscr{G}(\tau, \upsilon_0, \upsilon_1) = C_1$, $\mathscr{F}(\tau, \upsilon_0, \upsilon_1) = \mathscr{V}_4$, $\bar{\mathscr{F}}(\tau, \upsilon_0, \upsilon_1) = \mathscr{A}(2, 2, 2)$

Operator $F \in \bar{\mathscr{F}}$	$\Gamma^{(NCf)}\{\bar{\mathscr{F}}\}^a$	$\Gamma\{\bar{\mathscr{F}}\}^b$
E	$\begin{bmatrix} 1 & & & \\ & 1 & & \\ & & 1 & \\ & & & 1 \end{bmatrix} \otimes \begin{bmatrix} 1 & & \\ & 1 & \\ & & 1 \end{bmatrix}$	$\begin{bmatrix} 1 & . & . & . & . & . & . \\ & 1 & . & . & . & . & . \\ & & 1 & . & . & . & . \\ & & & 1 & . & . & . \\ & & & & 1 & . & .. \\ & & & & & 1 & . \\ & & & & & & 1 \end{bmatrix}$
F_2	$\begin{bmatrix} . & . & 1 & . \\ . & . & . & 1 \\ 1 & . & . & . \\ . & 1 & . & . \end{bmatrix} \otimes \begin{bmatrix} 1 & & \\ & -1 & \\ & & -1 \end{bmatrix}$	$\begin{bmatrix} 1 & . & . & . & . & . & \pi \\ & -1 & . & . & . & . & \pi \\ & & -1 & . & . & . & . \\ & & & 1 & . & . & . \\ & & & & . & 1 & . \\ & & & & 1 & . & . \\ & & & & & & 1 \end{bmatrix}$
F_3	$\begin{bmatrix} . & 1 & . & . \\ 1 & . & . & . \\ . & . & . & 1 \\ . & . & 1 & . \end{bmatrix} \otimes \begin{bmatrix} 1 & & \\ & -1 & \\ & & 1 \end{bmatrix}$	$\begin{bmatrix} 1 & . & . & . & . & . & \pi \\ & -1 & . & . & . & . & \pi \\ & & -1 & . & . & . & \pi \\ & & & -1 & . & . & . \\ & & & & -1 & . & . \\ & & & & & -1 & . \\ & & & & & & 1 \end{bmatrix}$

Table 9b (continued)

Operator $F \in \mathcal{F}$	$\Gamma^{(NCf)}\{\mathcal{F}\}^{a}$	$\Gamma\{\mathcal{F}\}^{b}$
F_4	$\begin{bmatrix} \cdot & \cdot & \cdot & 1 \\ \cdot & \cdot & 1 & \cdot \\ \cdot & 1 & \cdot & \cdot \\ 1 & \cdot & \cdot & \cdot \end{bmatrix} \otimes \begin{bmatrix} 1 & & \\ & 1 & \\ & & -1 \end{bmatrix}$	$\begin{bmatrix} 1 & \cdot & \cdot & \cdot & \cdot & \cdot & \cdot \\ & 1 & \cdot & \cdot & \cdot & \cdot & \cdot \\ & & 1 & \cdot & \cdot & \cdot & \pi \\ & & & -1 & \cdot & \cdot & \cdot \\ & & & & & -1 & \cdot \\ & & & & -1 & \cdot & \cdot \\ & & & & & & \cdot \end{bmatrix}$
$F_5 = F_p$	$\begin{bmatrix} 1 & \cdot & \cdot & \cdot \\ & 1 & \cdot & \cdot \\ & & 1 & \cdot \\ & & & 1 \end{bmatrix} \otimes \begin{bmatrix} -1 & & \\ & -1 & \\ & & 1 \end{bmatrix}$	$\begin{bmatrix} 1 & \cdot & \cdot & \cdot & \cdot & \cdot & \cdot \\ & 1 & \cdot & \cdot & \cdot & \cdot & \cdot \\ & & 1 & \cdot & \cdot & \cdot & \pi \\ & & & 1 & \cdot & \cdot & \pi \\ & & & & 1 & \cdot & \cdot \\ & & & & & 1 & \cdot \\ & & & & & & 1 \end{bmatrix}$
$F_6 = F_2 \cdot F_p$	$\begin{bmatrix} \cdot & \cdot & 1 & \cdot \\ \cdot & \cdot & \cdot & 1 \\ 1 & \cdot & \cdot & \cdot \\ \cdot & 1 & \cdot & \cdot \end{bmatrix} \otimes \begin{bmatrix} -1 & & \\ & 1 & \\ & & -1 \end{bmatrix}$	$\begin{bmatrix} 1 & \cdot & \cdot & \cdot & \cdot & \cdot & \pi \\ & -1 & \cdot & \cdot & \cdot & \cdot & \pi \\ & & -1 & \cdot & \cdot & \cdot & \pi \\ & & & 1 & \cdot & \cdot & \pi \\ & & & & \cdot & 1 & \cdot \\ & & & & 1 & \cdot & \cdot \\ & & & & & & 1 \end{bmatrix}$
$F_7 = F_3F_p$	$\begin{bmatrix} \cdot & 1 & \cdot & \cdot \\ 1 & \cdot & \cdot & \cdot \\ \cdot & \cdot & \cdot & 1 \\ \cdot & \cdot & 1 & \cdot \end{bmatrix} \otimes \begin{bmatrix} -1 & & \\ & 1 & \\ & & 1 \end{bmatrix}$	$\begin{bmatrix} 1 & \cdot & \cdot & \cdot & \cdot & \cdot & \pi \\ & -1 & \cdot & \cdot & \cdot & \cdot & \pi \\ & & -1 & \cdot & \cdot & \cdot & \cdot \\ & & & -1 & \cdot & \cdot & \pi \\ & & & & -1 & \cdot & \cdot \\ & & & & & -1 & \cdot \\ & & & & & & 1 \end{bmatrix}$
$F_8 = F_4F_p$	$\begin{bmatrix} \cdot & \cdot & \cdot & 1 \\ \cdot & \cdot & 1 & \cdot \\ \cdot & 1 & \cdot & \cdot \\ 1 & \cdot & \cdot & \cdot \end{bmatrix} \otimes \begin{bmatrix} -1 & & \\ & -1 & \\ & & -1 \end{bmatrix}$	$\begin{bmatrix} 1 & \cdot & \cdot & \cdot & \cdot & \cdot & \cdot \\ & 1 & \cdot & \cdot & \cdot & \cdot & \cdot \\ & & 1 & \cdot & \cdot & \cdot & \cdot \\ & & & -1 & \cdot & \cdot & \pi \\ & & & & \cdot & -1 & \cdot \\ & & & & -1 & \cdot & \cdot \\ & & & & & & 1 \end{bmatrix}$

[a] Representation of $\mathcal{F}$ generated by the vectors $\tilde{X}_{t00}(\tau, \upsilon_0)$, $\tilde{X}_{t01}(\tau, \upsilon_0)$, $\tilde{X}_{t10}(\tau, \upsilon_1)$, $\tilde{X}_{t11}(\tau, \upsilon_1)$ of a set of equivalent nuclei in general site originating from the two equivalent tops of local symmetry C_s (see Fig. 3)

[b] Representation of $\mathcal{F}$ by subsitutions of eulerian angles and internal coordinates:

$$\begin{bmatrix} \alpha' \\ \beta' \\ \gamma' \\ \tau' \\ \upsilon_0' \\ \upsilon_1' \\ 1 \end{bmatrix} = \begin{bmatrix} B(F) & 0 & b(F) \\ \cdot & A(F) & a(F) \\ \cdot & \cdot & 1 \end{bmatrix} \begin{bmatrix} \alpha \\ \beta \\ \gamma \\ \tau \\ \upsilon_0 \\ \upsilon_1 \\ 1 \end{bmatrix}, \qquad \begin{matrix} -\pi < \tau \leqslant +\pi, \ \tau \bmod 2p \\ -\pi < \upsilon_0, \upsilon_1 \leqslant +\pi \end{matrix}$$

$$\Gamma^{(NCf)}\{\overline{\mathscr{F}}\} \stackrel{is}{=} \mathscr{A}(2, 2, 2)$$

The rotative parts $\Gamma^{(3)}(F)$ form the point group

$$\Gamma^{(3)}\{\overline{\mathscr{K}}\} = \Gamma^{(3)}\{\overline{\mathscr{K}^+}\} \cup Z \cdot \Gamma^{(3)}\{\overline{\mathscr{K}^+}\} = D_{2h}$$

The proper rotative parts R(F) of the matrices $\Gamma^{(3)}(F)$ form the group

$$\Delta^{(3)}\{\overline{\mathscr{K}}\} \equiv \Gamma^{(3)}\{\overline{\mathscr{K}^+}\} := \left\{\begin{pmatrix}1&&\\&1&\\&&1\end{pmatrix}, \begin{pmatrix}1&&\\&-1&\\&&-1\end{pmatrix}, \begin{pmatrix}-1&&\\&1&\\&&-1\end{pmatrix}, \begin{pmatrix}-1&&\\&-1&\\&&1\end{pmatrix}\right\} = D_2$$

From this representation the substitution group of the eulerian angles

$$\mathscr{B}\{\overline{\mathscr{F}}\} := \left\{\begin{bmatrix}1&.&.&.\\&1&.&.\\&&1&.\\&&&1\end{bmatrix}, \begin{bmatrix}1&.&.&\pi\\&-1&.&\pi\\&&-1&.\\&&&1\end{bmatrix}, \begin{bmatrix}1&.&.&\pi\\&-1&.&\pi\\&&-1&\pi\\&&&1\end{bmatrix}, \begin{bmatrix}1&.&.&.\\&1&.&.\\&&1&\pi\\&&&1\end{bmatrix}\right\} = \mathscr{V}_4$$

and the representation

$$\Gamma\{\overline{\mathscr{F}}\} \stackrel{is}{=} \mathscr{A}(2, 2, 2)$$

is obtained along the procedure outlined in Sect. 2.2.1.1, cf. Table 9b.

Solution of Eq. (2.65) for all $F \in \overline{\mathscr{F}}$ gives the following fixed points:

F_2: $v_0 = v_1$, τ arbitrary : NC with covering symmetry $C_2(e_1^f)$

F_3: $v_0 = 0, \pi$, $\tau = 0$: NC with covering symmetry S_{13}^f

$v_1 = 0, \pi$

F_4: $v_0 = -v_1$, $\tau = 0$: NC with covering symmetry S_{12}^f

If a NC is a fixed point NC of F_2, F_3 and F_4 simultaneously, e.g. with $v_0 = v_1 = \tau = 0$ or $v_0 = v_1 = \pi$, $\tau = 0$, it possesses covering symmetry C_{2v}.

F_5 and F_6 do not have fixed points, but

F_7: $v_0 = 0, \pi$, $\tau = \pi/2$: NC with covering symmetry S_{23}^f

$v_1 = 0, \pi$

F_8: $v_0 = -v_1$, $\tau = \pi/2$: NC with covering symmetry Z

Fixed point NCs of F_2, F_7 and F_8 ($v_0 = v_1 = 0$, $\tau = \pi/2$ or $v_0 = v_1 = \pi$, $\tau = \pi/2$) have covering symmetry C_{2h}.

This example shows how conformations with specially high covering symmetry may systematically be derived by means of the isometric group.

The Longuet-Higgins group $\mathscr{L}\mathscr{H}$ of ethylene glycol type molecules is endomorphic to $\overline{\mathscr{F}}(\tau, v_0, v_1)$ because

$$\Gamma^{(NCl)}(F_p) = 1^{(4)} \otimes |\Gamma^{(3)}(F_p)| \cdot 1^{(3)} = 1^{(4)} \otimes 1^{(3)},$$

$$\Gamma^{(NCl)}\{\overline{\mathscr{F}}\} := \{1^{(4)} \otimes 1^{(3)}, \Pi(F_2) \otimes 1^{(3)}, \Pi(F_3) \otimes (-1^{(3)}), \Pi(F_4) \otimes (-1^{(3)})\}$$

$$\mathscr{L}\mathscr{H} \overset{is}{=} \Gamma^{(NCl)}\{\overline{\mathscr{F}}\} \overset{is}{=} \mathscr{V}_4 \overset{ho}{=} \overline{\mathscr{F}}(\tau, \upsilon_0, \upsilon_1)$$

From the isometric group of the $C_2(\tau)F(C_sT)_2$ system the symmetry groups of a considerable number of further SRMs may be obtained, e.g.:

(i) freezing τ at 0 gives a $C_{2v}F(C_sT)_2$ system like 1,2-dihydroxy benzene;

(ii) freezing τ at π leads to a $C_{2h}F(C_sT)_2$ system, e.g. *trans*-1,4-dichlorobutene-2;

(iii) if both τ and υ_0 are kept constant at 0 or π, we have a $C_sF\ C_sT$ system like acrolein;

(iv) if both υ_0 and υ_1 are frozen either at 0 or at π, we have a system with two equivalent C_s tops, $D_{\infty h}F(C_sT)_2$, for which glyoxal or 1,2-difluoroethane are examples.

2.4.3 $D_2(\tau)F(C_1TR)(C_1TS)$ and $D_2(\tau)F(C_1TR)_2$ Systems

Molecules consisting of a semirigid frame with covering symmetry D_2, to which two rigid tops of local symmetry C_1 with opposite or equal configuration are attached ($D_2(\tau)F(C_1TR)(C_1TS)$ and $D_2(\tau)F(C_1TR)_2$, respectively) are interesting examples w.r.t. the chirality problem of nonrigid molecules (see Sect. 3.4). A molecule of type $D_2(\tau)F(C_1TR)(C_1TS)$ is scetched in Fig. 5a. Figure 5b shows the definition of the frame system and the internal coordinates. Table 10 gives the coordinate vectors of sets of equivalent frame and top nuclei.

(i) $D_2(\tau)F(C_1TR)(C_1TS)$ System

The following formula holds for distances between top nuclei with local coordinates $\widetilde{X}^t_{t0} = (X^t_{t01}\ 0\ X^t_{t03})$ of the representative, e.g. the Cl nuclei, and frame nuclei denoted by Z in Fig. 5

$$\begin{aligned} d^2_{t\lambda, f\overline{\lambda}\mu}(\tau, \upsilon_\lambda) &= |X_{t\lambda}(\tau, \upsilon_\lambda) - X_{f\overline{\lambda}\mu}(\tau)|^2 \\ &= \frac{1}{4} r^2 + r \cdot X^t_{t03} + \widetilde{X}^t_{t0} X^t_{t0} + \widetilde{X}^f_{f00} X^f_{f00} + \\ &+ (-1)^{\lambda+\overline{\lambda}+1} r \cdot X^f_{f003} + 2(-1)^{\lambda+\overline{\lambda}+1} X^t_{t03} X^f_{f003} + \\ &+ (-1)^{\mu+1} X^f_{f001} X^t_{t01} [c\upsilon_\lambda(1 + c2\tau) - s\upsilon_\lambda s2\tau] + \\ &+ (-1)^{\lambda+\overline{\lambda}+\mu+1} X^f_{f001} X^t_{t01} [c\upsilon_\lambda(1 - c2\tau) + s\upsilon_\lambda s2\tau] \\ &\quad (\lambda, \overline{\lambda}, \mu = n \pmod 2) \end{aligned}$$

Thus the primitive period of τ is π, that of υ_0 and υ_1 is 2π. The internal isometric transformations $\mathscr{A}$(F) for the domains $-\pi/2 < \tau \leqslant +\pi/2$ (τ mod p), $-\pi < \upsilon_0$, $\upsilon_1 \leqslant +\pi$ are listed in Table 11a and may be obtained either from the distance func-

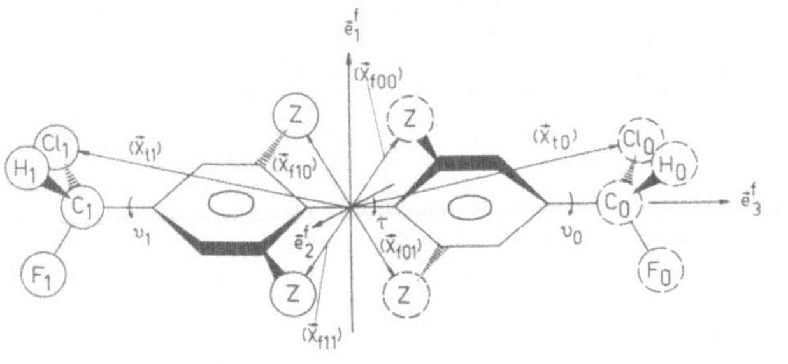

a

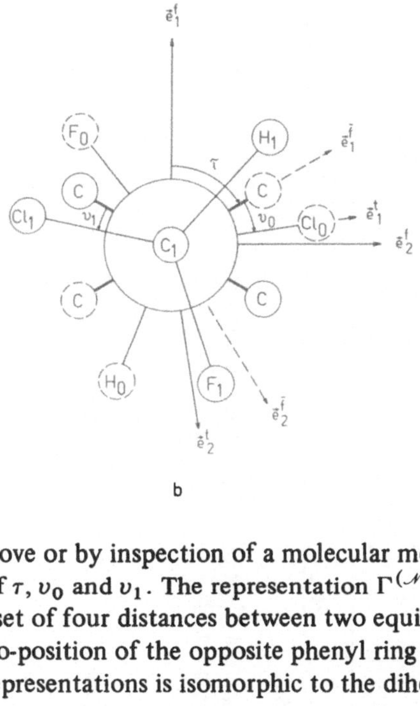

b

Fig. 5. $D_2(\tau)F(C_1TR)(C_1TS)$ system.
(a) Pictorial view, **(b)** Newman projection (symbolic).
Key: The origin of the local top coordinate system $\tilde{e}^t$ is situated in the nucleus C_0: the origin of the local coordinate system $\tilde{e}^f$ coincides with the origin of the frame system $\tilde{e}^f$

tion above or by inspection of a molecular model for distance preserving transformations of τ, υ_0 and υ_1. The representation $\Gamma^{(\mathcal{N}\mathcal{C})}\{\mathcal{F}\}$ of $\mathcal{F}$ by permutations refers to the set of four distances between two equivalent top nuclei and frame nuclei Z in ortho-position of the opposite phenyl ring (cf. Fig. 5a). The abstract group $\mathcal{F}$ of both representations is isomorphic to the dihedral group of order 8

$$\mathcal{F}(\tau, \upsilon_0, \upsilon_1) \overset{\text{is}}{=} \mathcal{A}\{\mathcal{F}\} \overset{\text{is}}{=} \Gamma^{(\mathcal{N}\mathcal{C})}\{\mathcal{F}\} \overset{\text{is}}{=} \vartheta_4$$

with $\mathcal{F} = \{F_2^k\} \cup F_5\{F_2^k\}$, $F_2^kF_5 = F_5F_2^{-k}$

$\tau' = \tau + \pi$ is a nontrivial primitive period operation: application of this substitution to the coordinate vectors of two equivalent top and four equivalent frame nuclei gives

$$\hat{P}_{F_p}\{\widetilde{X}_{t\lambda}(\tau, \upsilon_\lambda)\widetilde{X}_{f\bar{\lambda}\mu}(\tau)\} = \{\widetilde{X}_{t\lambda}(\tau - \pi, \upsilon_\lambda)\widetilde{X}_{f\bar{\lambda}\mu}(\tau - \pi)\}$$

$$= \{\widetilde{X}_{t\lambda}(\tau, \upsilon_\lambda)\widetilde{X}_{f\bar{\lambda}\mu}(\tau)\}\, 1^{(6)} \otimes \begin{bmatrix} -1 & & \\ & -1 & \\ & & 1 \end{bmatrix}$$

$$\lambda, \bar{\lambda}, \mu = 0, 1$$

i.e. $\Gamma^{(3)}(F_p) \neq 1^{(3)}$, $\Gamma^{(\mathcal{N}\mathcal{C})}(F_p) = 1^{(4)}$

Table 10. Coordinate vectors of equivalent nuclei of $D_2(\tau)F(C_1TR)_2$ and $D_2(\tau)F(C_1TR)(C_1TS)$ systems

Set	Vector[a]	X
Frame $\bar{\lambda}, \mu = 0,1$	$X_{f\bar{\lambda}\mu}$	$\begin{bmatrix} 1 & \cdot & \cdot \\ \cdot & -1 & \cdot \\ \cdot & \cdot & -1 \end{bmatrix}^{\bar{\lambda}} \cdot \begin{bmatrix} -1 & \cdot & \cdot \\ \cdot & -1 & \cdot \\ \cdot & \cdot & 1 \end{bmatrix}^{\mu} \cdot \begin{bmatrix} c\tau & -s\tau & 0 \\ s\tau & c\tau & 0 \\ 0 & 0 & 1 \end{bmatrix} X^{\bar{f}}_{f00}$ [b]
$D_2(\tau)F(C_1TR)_2$ system Top λ $\lambda = 0,1$	$X_{t\lambda}$	$\begin{bmatrix} 1 & \cdot & \cdot \\ \cdot & -1 & \cdot \\ \cdot & \cdot & -1 \end{bmatrix}^{\lambda} \left\{ \frac{1}{2} r \begin{bmatrix} 0 \\ 0 \\ 1 \end{bmatrix} + \begin{bmatrix} c\tau & -s\tau & 0 \\ s\tau & c\tau & 0 \\ 0 & 0 & 1 \end{bmatrix} \cdot \begin{bmatrix} cv\lambda & -sv\lambda & 0 \\ sv\lambda & cv\lambda & 0 \\ 0 & 0 & 1 \end{bmatrix} X^{t}_{t0} \text{ [b]} \right\}$
$D_2(\tau)F(C_1TR)(C_1TS)$ system	$X_{t\lambda}$	$\begin{bmatrix} 1 & & \\ & -1 & \\ & & -1 \end{bmatrix}^{\lambda} \left\{ \frac{1}{2} r \begin{bmatrix} 0 \\ 0 \\ 1 \end{bmatrix} + \begin{bmatrix} c\tau & -s\tau & 0 \\ s\tau & c\tau & 0 \\ 0 & 0 & 1 \end{bmatrix} \begin{bmatrix} cv\lambda & -sv\lambda & 0 \\ sv\lambda & cv\lambda & 0 \\ 0 & 0 & 1 \end{bmatrix} \begin{bmatrix} 1 & & \\ & -1 & \\ & & 1 \end{bmatrix}^{\lambda} X^{t}_{t0} \text{ [b]} \right\}$

[a] Coordinates refer to the frame coordinate system $\tilde{e}^f$ indicated in Fig. 5; r is the distance between the carbon nuclei of the two substituted methyl groups (see Fig. 5a).

[b] Coordinate vectors $\tilde{X}^{\bar{f}}_{f00} = (X^{\bar{f}}_{f001}\, 0\, X^{\bar{f}}_{f003})$ and $X^t_{t0} = (X^t_{t01}\, X^t_{t02}\, X^t_{t03})$ of representative nuclei w.r.t. local coordinate systems $\tilde{e}^f$ and $\tilde{e}^t$, respectively, indicated in Fig. 5b.

Table 11a. Isometric group of the $D_2(\tau)F(C_1TR)(C_1TS)$ system
$\mathcal{G}(\tau, v_0, v_1) \overset{\text{is}}{=} \mathcal{C}_1$, $\mathcal{F}(\tau, v_0, v_1) \overset{\text{is}}{=} \vartheta_4$, $\overline{\mathcal{F}}(\tau, v_0, v_1) \overset{\text{is}}{=} \mathcal{G}_{16}$

Operator	$\mathcal{A}\{\mathcal{F}\}$[a]	$\Gamma^{(\mathcal{N}\mathcal{C})}\{\mathcal{F}\}$[b]	Operator	$\mathcal{A}\{\mathcal{F}\}$	$\Gamma^{(\mathcal{N}\mathcal{C})}\{\mathcal{F}\}$
E	$\begin{bmatrix} 1 & \cdot & \cdot & \cdot \\ \cdot & 1 & \cdot & \cdot \\ \cdot & \cdot & 1 & \cdot \\ \cdot & \cdot & \cdot & 1 \end{bmatrix}$	$\begin{bmatrix} 1 & \cdot & \cdot & \cdot \\ \cdot & 1 & \cdot & \cdot \\ \cdot & \cdot & 1 & \cdot \\ \cdot & \cdot & \cdot & 1 \end{bmatrix}$	F_5	$\begin{bmatrix} \cdot & -1 & \cdot & \cdot \\ -1 & \cdot & \cdot & \cdot \\ \cdot & \cdot & -1 & \cdot \\ \cdot & \cdot & \cdot & 1 \end{bmatrix}$	$\begin{bmatrix} \cdot & \cdot & 1 & \cdot \\ \cdot & \cdot & \cdot & 1 \\ 1 & \cdot & \cdot & \cdot \\ \cdot & 1 & \cdot & \cdot \end{bmatrix}$
F_2	$\begin{bmatrix} \cdot & -1 & \cdot & \\ -1 & \cdot & \cdot & \pi \\ \cdot & \cdot & -1 & \pi/2 \\ \cdot & \cdot & \cdot & 1 \end{bmatrix}$	$\begin{bmatrix} \cdot & \cdot & \cdot & 1 \\ \cdot & \cdot & 1 & \cdot \\ 1 & \cdot & \cdot & \cdot \\ \cdot & 1 & \cdot & \cdot \end{bmatrix}$	F_6	$\begin{bmatrix} 1 & \cdot & \cdot & \cdot \\ \cdot & 1 & \cdot & \pi \\ \cdot & \cdot & 1 & \pi/2 \\ \cdot & \cdot & \cdot & 1 \end{bmatrix}$	$\begin{bmatrix} \cdot & 1 & \cdot & \cdot \\ 1 & \cdot & \cdot & \cdot \\ \cdot & \cdot & 1 & \cdot \\ \cdot & \cdot & \cdot & 1 \end{bmatrix}$
F_3	$\begin{bmatrix} 1 & \cdot & \cdot & \pi \\ \cdot & 1 & \cdot & \pi \\ \cdot & \cdot & 1 & \cdot \\ \cdot & \cdot & \cdot & 1 \end{bmatrix}$	$\begin{bmatrix} \cdot & 1 & \cdot & \cdot \\ 1 & \cdot & \cdot & \cdot \\ \cdot & \cdot & \cdot & 1 \\ \cdot & \cdot & 1 & \cdot \end{bmatrix}$	F_7	$\begin{bmatrix} \cdot & -1 & \cdot & \pi \\ -1 & \cdot & \cdot & \pi \\ \cdot & \cdot & -1 & \cdot \\ \cdot & \cdot & \cdot & 1 \end{bmatrix}$	$\begin{bmatrix} \cdot & \cdot & \cdot & 1 \\ \cdot & \cdot & 1 & \cdot \\ \cdot & 1 & \cdot & \cdot \\ 1 & \cdot & \cdot & \cdot \end{bmatrix}$
F_4	$\begin{bmatrix} \cdot & -1 & \cdot & \pi \\ -1 & \cdot & \cdot & \cdot \\ \cdot & \cdot & -1 & \pi/2 \\ \cdot & \cdot & \cdot & 1 \end{bmatrix}$	$\begin{bmatrix} \cdot & \cdot & 1 & \cdot \\ \cdot & \cdot & \cdot & 1 \\ \cdot & 1 & \cdot & \cdot \\ 1 & \cdot & \cdot & \cdot \end{bmatrix}$	F_8	$\begin{bmatrix} 1 & \cdot & \cdot & \pi \\ \cdot & 1 & \cdot & \cdot \\ \cdot & \cdot & 1 & \pi/2 \\ \cdot & \cdot & \cdot & 1 \end{bmatrix}$	$\begin{bmatrix} 1 & \cdot & \cdot & \cdot \\ \cdot & 1 & \cdot & \cdot \\ \cdot & \cdot & \cdot & 1 \\ \cdot & \cdot & 1 & \cdot \end{bmatrix}$

[a] Representation of $\mathcal{F}$ by substitutions of the internal coordinates:

$$\begin{bmatrix} v_0' \\ v_1' \\ \tau' \\ 1 \end{bmatrix} = \begin{bmatrix} A(F) & a(F) \\ 0 & 1 \end{bmatrix} \cdot \begin{bmatrix} v_0 \\ v_1 \\ \tau \\ 1 \end{bmatrix} \qquad \begin{matrix} -\pi/2 < \tau \leqslant +\pi/2, \tau \text{ mod p} \\ -\pi < v_0, v_1 \leqslant +\pi \end{matrix}$$

[b] Representation by permutations generated by the set of distances $d_{t0,f10}(v_0, \tau)$, $d_{t0,f11}(v_0, \tau)$, $d_{t1,f00}(v_1, \tau)$, $d_{t1,f01}(v_1, \tau)$ (see Fig. 5a).

Table 11b. Isometric group of the $D_2(\tau)F(C_1TR)(C_1TS)$ system: $\mathscr{G}(\tau, v_0, v_1) \overset{is}{\equiv} C_1$, $\mathscr{F}(\tau, v_0, v_1) \overset{is}{\equiv} \vartheta_4$, $\overline{\mathscr{F}}(\tau, v_0, v_1) \overset{is}{\equiv} \mathscr{G}_{16}$

Operator $F \in \overline{\mathscr{F}}$	$\Gamma^{(NCf)}\{\overline{\mathscr{F}}\}^a$	$\Gamma\{\overline{\mathscr{F}}\}^b$
$E, T = F_p$	$\begin{bmatrix}1&\cdot&\cdot&\cdot&\cdot&\cdot\\\cdot&1&\cdot&\cdot&\cdot&\cdot\\\cdot&\cdot&1&\cdot&\cdot&\cdot\\\cdot&\cdot&\cdot&1&\cdot&\cdot\\\cdot&\cdot&\cdot&\cdot&1&\cdot\\\cdot&\cdot&\cdot&\cdot&\cdot&1\end{bmatrix} \otimes \begin{bmatrix}1&\cdot&\cdot\\\cdot&1&\cdot\\\cdot&\cdot&1\end{bmatrix}, \begin{bmatrix}1&\cdot&\cdot&\cdot&\cdot&\cdot\\\cdot&1&\cdot&\cdot&\cdot&\cdot\\\cdot&\cdot&1&\cdot&\cdot&\cdot\\\cdot&\cdot&\cdot&1&\cdot&\cdot\\\cdot&\cdot&\cdot&\cdot&1&\cdot\\\cdot&\cdot&\cdot&\cdot&\cdot&1\end{bmatrix} \otimes \begin{bmatrix}-1&\cdot&\cdot\\\cdot&-1&\cdot\\\cdot&\cdot&1\end{bmatrix}$	$\begin{bmatrix}1&\cdot&\cdot&\cdot&\cdot&\cdot&\cdot\\&1&\cdot&\cdot&\cdot&\cdot&\cdot\\&&1&\cdot&\cdot&\cdot&\cdot\\&&&1&\cdot&\cdot&\cdot\\&&&&1&\cdot&\cdot\\&&&&&1&\cdot\\&&&&&&1\end{bmatrix} \begin{bmatrix}1&\cdot&\cdot&\cdot&\cdot&\cdot&\cdot\\&1&\cdot&\cdot&\cdot&\cdot&\cdot\\&&1&\cdot&\cdot&\cdot&\pi\\&&&1&\cdot&\cdot&\cdot\\&&&&1&\cdot&\cdot\\&&&&&1&\pi\\&&&&&&1\end{bmatrix}$
C, TC	$\begin{bmatrix}\cdot&1&\cdot&\cdot&\cdot&\cdot\\1&\cdot&\cdot&\cdot&\cdot&\cdot\\\cdot&\cdot&\cdot&\cdot&1&\cdot\\\cdot&\cdot&\cdot&\cdot&\cdot&1\\\cdot&\cdot&\cdot&1&\cdot&\cdot\\\cdot&\cdot&1&\cdot&\cdot&\cdot\end{bmatrix} \otimes \begin{bmatrix}\cdot&-1&\cdot\\1&\cdot&\cdot\\\cdot&\cdot&-1\end{bmatrix}, \begin{bmatrix}\cdot&1&\cdot&\cdot&\cdot&\cdot\\1&\cdot&\cdot&\cdot&\cdot&\cdot\\\cdot&\cdot&\cdot&\cdot&1&\cdot\\\cdot&\cdot&\cdot&\cdot&\cdot&1\\\cdot&\cdot&\cdot&1&\cdot&\cdot\\\cdot&\cdot&1&\cdot&\cdot&\cdot\end{bmatrix} \otimes \begin{bmatrix}\cdot&1&\cdot\\-1&\cdot&\cdot\\\cdot&\cdot&-1\end{bmatrix}$	$\begin{bmatrix}1&\cdot&\cdot&\cdot&\cdot&\cdot&\cdot\\&1&\cdot&\cdot&\cdot&\cdot&\cdot\\&&1&\cdot&\cdot&\cdot&\pi/2\\&&&&-1&\cdot&\cdot\\&&&-1&\cdot&\cdot&\pi\\&&&&\cdot&-1&\pi/2\\&&&&&&1\end{bmatrix} \begin{bmatrix}1&\cdot&\cdot&\cdot&\cdot&\cdot&\cdot\\&1&\cdot&\cdot&\cdot&\cdot&\cdot\\&&1&\cdot&\cdot&\cdot&-\pi/2\\&&&\cdot&-1&\cdot&\cdot\\&&&-1&\cdot&\cdot&\pi\\&&&&&-1&-\pi/2\\&&&&&&1\end{bmatrix}$
C^2, TC^2	$\begin{bmatrix}1&\cdot&\cdot&\cdot&\cdot&\cdot\\\cdot&1&\cdot&\cdot&\cdot&\cdot\\\cdot&\cdot&\cdot&1&\cdot&\cdot\\\cdot&\cdot&1&\cdot&\cdot&\cdot\\\cdot&\cdot&\cdot&\cdot&\cdot&1\\\cdot&\cdot&\cdot&\cdot&1&\cdot\end{bmatrix} \otimes \begin{bmatrix}-1&\cdot&\cdot\\\cdot&-1&\cdot\\\cdot&\cdot&1\end{bmatrix}, \begin{bmatrix}1&\cdot&\cdot&\cdot&\cdot&\cdot\\\cdot&1&\cdot&\cdot&\cdot&\cdot\\\cdot&\cdot&\cdot&1&\cdot&\cdot\\\cdot&\cdot&1&\cdot&\cdot&\cdot\\\cdot&\cdot&\cdot&\cdot&\cdot&1\\\cdot&\cdot&\cdot&\cdot&1&\cdot\end{bmatrix} \otimes \begin{bmatrix}1&\cdot&\cdot\\\cdot&1&\cdot\\\cdot&\cdot&1\end{bmatrix}$	$\begin{bmatrix}1&\cdot&\cdot&\cdot&\cdot&\cdot&\cdot\\&1&\cdot&\cdot&\cdot&\cdot&\cdot\\&&1&\cdot&\cdot&\cdot&\pi\\&&&1&\cdot&\cdot&\pi\\&&&&1&\cdot&\pi\\&&&&&1&\cdot\\&&&&&&1\end{bmatrix} \begin{bmatrix}1&\cdot&\cdot&\cdot&\cdot&\cdot&\cdot\\&1&\cdot&\cdot&\cdot&\cdot&\cdot\\&&1&\cdot&\cdot&\cdot&\cdot\\&&&1&\cdot&\cdot&\pi\\&&&&1&\cdot&\pi\\&&&&&1&\pi\\&&&&&&1\end{bmatrix}$
C^3, TC^3	$\begin{bmatrix}\cdot&1&\cdot&\cdot&\cdot&\cdot\\1&\cdot&\cdot&\cdot&\cdot&\cdot\\\cdot&\cdot&\cdot&\cdot&\cdot&1\\\cdot&\cdot&\cdot&\cdot&1&\cdot\\\cdot&\cdot&1&\cdot&\cdot&\cdot\\\cdot&\cdot&\cdot&1&\cdot&\cdot\end{bmatrix} \otimes \begin{bmatrix}\cdot&1&\cdot\\-1&\cdot&\cdot\\\cdot&\cdot&-1\end{bmatrix}, \begin{bmatrix}\cdot&1&\cdot&\cdot&\cdot&\cdot\\1&\cdot&\cdot&\cdot&\cdot&\cdot\\\cdot&\cdot&\cdot&\cdot&\cdot&1\\\cdot&\cdot&\cdot&\cdot&1&\cdot\\\cdot&\cdot&1&\cdot&\cdot&\cdot\\\cdot&\cdot&\cdot&1&\cdot&\cdot\end{bmatrix} \otimes \begin{bmatrix}\cdot&-1&\cdot\\1&\cdot&\cdot\\\cdot&\cdot&-1\end{bmatrix}$	$\begin{bmatrix}1&\cdot&\cdot&\cdot&\cdot&\cdot&\cdot\\&1&\cdot&\cdot&\cdot&\cdot&\cdot\\&&1&\cdot&\cdot&\cdot&-\pi/2\\&&&&-1&\cdot&\pi\\&&&-1&\cdot&\cdot&\cdot\\&&&&&-1&\pi/2\\&&&&&&1\end{bmatrix} \begin{bmatrix}1&\cdot&\cdot&\cdot&\cdot&\cdot&\cdot\\&1&\cdot&\cdot&\cdot&\cdot&\cdot\\&&1&\cdot&\cdot&\cdot&\pi/2\\&&&&-1&\cdot&\pi\\&&&-1&\cdot&\cdot&\cdot\\&&&&&-1&-\pi/2\\&&&&&&1\end{bmatrix}$

U, TU

$$\begin{bmatrix} \cdot & 1 & \cdot & \cdot & \cdot & \cdot \\ 1 & \cdot & \cdot & \cdot & \cdot & \cdot \\ \cdot & \cdot & \cdot & \cdot & 1 & \cdot \\ \cdot & \cdot & \cdot & \cdot & \cdot & 1 \\ \cdot & \cdot & 1 & \cdot & \cdot & \cdot \\ \cdot & \cdot & \cdot & 1 & \cdot & \cdot \end{bmatrix} \otimes \begin{bmatrix} 1 & \cdot & \cdot \\ \cdot & 1 & \cdot \\ \cdot & \cdot & -1 \end{bmatrix}, \begin{bmatrix} \cdot & 1 & \cdot & \cdot & \cdot & \cdot \\ 1 & \cdot & \cdot & \cdot & \cdot & \cdot \\ \cdot & \cdot & \cdot & \cdot & 1 & \cdot \\ \cdot & \cdot & \cdot & \cdot & \cdot & 1 \\ \cdot & \cdot & 1 & \cdot & \cdot & \cdot \\ \cdot & \cdot & \cdot & 1 & \cdot & \cdot \end{bmatrix} \otimes \begin{bmatrix} -1 & \cdot & \cdot \\ \cdot & -1 & \cdot \\ \cdot & \cdot & -1 \end{bmatrix} \quad \begin{bmatrix} 1 & \cdot & \cdot & \cdot & \cdot & \cdot & \cdot \\ & 1 & \cdot & \cdot & \cdot & \cdot & \cdot \\ & & 1 & \cdot & \cdot & \cdot & \pi \\ & & & \cdot & -1 & \cdot & \cdot \\ & & & -1 & \cdot & \cdot & \cdot \\ & & & & & -1 & \cdot \\ & & & & & & 1 \end{bmatrix} \quad \begin{bmatrix} 1 & \cdot & \cdot & \cdot & \cdot & \cdot & \cdot \\ & 1 & \cdot & \cdot & \cdot & \cdot & \cdot \\ & & 1 & \cdot & \cdot & \cdot & \cdot \\ & & & \cdot & -1 & \cdot & \cdot \\ & & & -1 & \cdot & \cdot & \cdot \\ & & & & & -1 & \pi \\ & & & & & & 1 \end{bmatrix}$$

CU, TCU

$$\begin{bmatrix} 1 & \cdot & \cdot & \cdot & \cdot & \cdot \\ \cdot & 1 & \cdot & \cdot & \cdot & \cdot \\ \cdot & \cdot & 1 & \cdot & \cdot & \cdot \\ \cdot & \cdot & \cdot & 1 & \cdot & \cdot \\ \cdot & \cdot & \cdot & \cdot & \cdot & 1 \\ \cdot & \cdot & \cdot & \cdot & 1 & \cdot \end{bmatrix} \otimes \begin{bmatrix} \cdot & -1 & \cdot \\ 1 & \cdot & \cdot \\ \cdot & \cdot & 1 \end{bmatrix}, \begin{bmatrix} 1 & \cdot & \cdot & \cdot & \cdot & \cdot \\ \cdot & 1 & \cdot & \cdot & \cdot & \cdot \\ \cdot & \cdot & 1 & \cdot & \cdot & \cdot \\ \cdot & \cdot & \cdot & 1 & \cdot & \cdot \\ \cdot & \cdot & \cdot & \cdot & \cdot & 1 \\ \cdot & \cdot & \cdot & \cdot & 1 & \cdot \end{bmatrix} \otimes \begin{bmatrix} \cdot & 1 & \cdot \\ -1 & \cdot & \cdot \\ \cdot & \cdot & 1 \end{bmatrix} \quad \begin{bmatrix} 1 & \cdot & \cdot & \cdot & \cdot & \cdot & \cdot \\ & 1 & \cdot & \cdot & \cdot & \cdot & \cdot \\ & & 1 & \cdot & \cdot & \cdot & -\pi/2 \\ & & & 1 & \cdot & \cdot & \cdot \\ & & & & 1 & \cdot & \pi \\ & & & & & 1 & \pi/2 \\ & & & & & & 1 \end{bmatrix} \quad \begin{bmatrix} 1 & \cdot & \cdot & \cdot & \cdot & \cdot & \cdot \\ & 1 & \cdot & \cdot & \cdot & \cdot & \cdot \\ & & 1 & \cdot & \cdot & \cdot & \pi/2 \\ & & & 1 & \cdot & \cdot & \cdot \\ & & & & 1 & \cdot & \pi \\ & & & & & 1 & -\pi/2 \\ & & & & & & 1 \end{bmatrix}$$

C^2U, TC^2U

$$\begin{bmatrix} \cdot & 1 & \cdot & \cdot & \cdot & \cdot \\ 1 & \cdot & \cdot & \cdot & \cdot & \cdot \\ \cdot & \cdot & \cdot & \cdot & \cdot & 1 \\ \cdot & \cdot & \cdot & \cdot & 1 & \cdot \\ \cdot & \cdot & \cdot & 1 & \cdot & \cdot \\ \cdot & \cdot & 1 & \cdot & \cdot & \cdot \end{bmatrix} \otimes \begin{bmatrix} -1 & \cdot & \cdot \\ \cdot & -1 & \cdot \\ \cdot & \cdot & -1 \end{bmatrix}, \begin{bmatrix} \cdot & 1 & \cdot & \cdot & \cdot & \cdot \\ 1 & \cdot & \cdot & \cdot & \cdot & \cdot \\ \cdot & \cdot & \cdot & \cdot & \cdot & 1 \\ \cdot & \cdot & \cdot & \cdot & 1 & \cdot \\ \cdot & \cdot & \cdot & 1 & \cdot & \cdot \\ \cdot & \cdot & 1 & \cdot & \cdot & \cdot \end{bmatrix} \otimes \begin{bmatrix} 1 & \cdot & \cdot \\ \cdot & 1 & \cdot \\ \cdot & \cdot & -1 \end{bmatrix} \quad \begin{bmatrix} 1 & \cdot & \cdot & \cdot & \cdot & \cdot & \cdot \\ & 1 & \cdot & \cdot & \cdot & \cdot & \cdot \\ & & 1 & \cdot & \cdot & \cdot & \cdot \\ & & & \cdot & -1 & \cdot & \pi \\ & & & -1 & \cdot & \cdot & \pi \\ & & & & & -1 & \cdot \\ & & & & & & 1 \end{bmatrix} \quad \begin{bmatrix} 1 & \cdot & \cdot & \cdot & \cdot & \cdot & \cdot \\ & 1 & \cdot & \cdot & \cdot & \cdot & \cdot \\ & & 1 & \cdot & \cdot & \cdot & \pi \\ & & & \cdot & -1 & \cdot & \pi \\ & & & -1 & \cdot & \cdot & \pi \\ & & & & & -1 & \pi \\ & & & & & & 1 \end{bmatrix}$$

C^3U, TC^3U

$$\begin{bmatrix} 1 & \cdot & \cdot & \cdot & \cdot & \cdot \\ \cdot & 1 & \cdot & \cdot & \cdot & \cdot \\ \cdot & \cdot & \cdot & 1 & \cdot & \cdot \\ \cdot & \cdot & 1 & \cdot & \cdot & \cdot \\ \cdot & \cdot & \cdot & \cdot & 1 & \cdot \\ \cdot & \cdot & \cdot & \cdot & \cdot & 1 \end{bmatrix} \otimes \begin{bmatrix} \cdot & 1 & \cdot \\ -1 & \cdot & \cdot \\ \cdot & \cdot & 1 \end{bmatrix}, \begin{bmatrix} 1 & \cdot & \cdot & \cdot & \cdot & \cdot \\ \cdot & 1 & \cdot & \cdot & \cdot & \cdot \\ \cdot & \cdot & \cdot & 1 & \cdot & \cdot \\ \cdot & \cdot & 1 & \cdot & \cdot & \cdot \\ \cdot & \cdot & \cdot & \cdot & 1 & \cdot \\ \cdot & \cdot & \cdot & \cdot & \cdot & 1 \end{bmatrix} \otimes \begin{bmatrix} \cdot & -1 & \cdot \\ 1 & \cdot & \cdot \\ \cdot & \cdot & 1 \end{bmatrix} \quad \begin{bmatrix} 1 & \cdot & \cdot & \cdot & \cdot & \cdot & \cdot \\ & 1 & \cdot & \cdot & \cdot & \cdot & \cdot \\ & & 1 & \cdot & \cdot & \cdot & +\pi/2 \\ & & & 1 & \cdot & \cdot & \pi \\ & & & & 1 & \cdot & \cdot \\ & & & & & 1 & \pi/2 \\ & & & & & & 1 \end{bmatrix} \quad \begin{bmatrix} 1 & \cdot & \cdot & \cdot & \cdot & \cdot & \cdot \\ & 1 & \cdot & \cdot & \cdot & \cdot & \cdot \\ & & 1 & \cdot & \cdot & \cdot & -\pi/2 \\ & & & 1 & \cdot & \cdot & \pi \\ & & & & 1 & \cdot & \cdot \\ & & & & & 1 & -\pi/2 \\ & & & & & & 1 \end{bmatrix}$$

---

[a] Representation generated by the vectors $\widetilde{X}_{t0}(\upsilon_0, \tau)$, $\widetilde{X}_{t1}(\upsilon_1, \tau)$, $\widetilde{X}_{f00}(\tau)$, $\widetilde{X}_{f01}(\tau)$, $\widetilde{X}_{f10}(\tau)$, $\widetilde{X}_{f11}(\tau)$ of two sets of equivalent nuclei originating from the two equivalent tops and the frame, respectively (see Fig. 5a).

[b] Representation of $\overline{\mathscr{F}}$ by substitutions of eulerian angles and internal coordinates:

$$\begin{bmatrix} \alpha' \\ \beta' \\ \gamma' \\ \upsilon_0' \\ \upsilon_1' \\ \tau' \\ 1 \end{bmatrix} = \begin{bmatrix} B(F) & 0 & b(F) \\ \cdot & A(F) & a(F) \\ \cdot & \cdot & 1 \end{bmatrix} \cdot \begin{bmatrix} \alpha \\ \beta \\ \gamma \\ \upsilon_0 \\ \upsilon_1 \\ \tau \\ 1 \end{bmatrix}, \quad \begin{matrix} -\pi < \tau \leqslant +\pi, \tau \bmod 2\,p \\ -\pi < \upsilon_0, \upsilon_1 \leqslant \pi \end{matrix}$$

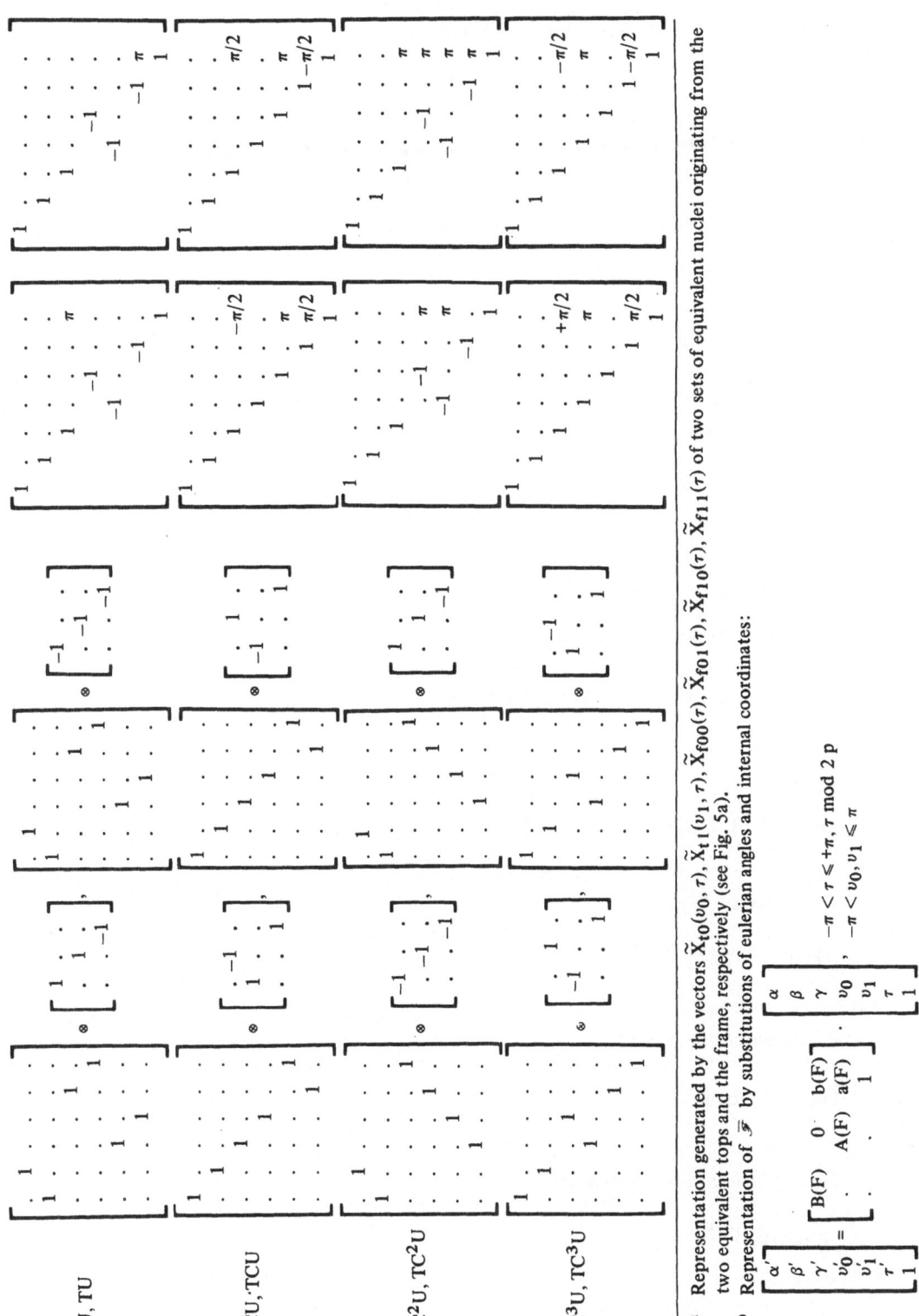

Taking τ modulo 2 p = 2 π, i.e. admitting

$$\mathscr{A}(F_p) = \begin{bmatrix} 1 & . & . & . \\ & 1 & . & . \\ & & 1 & \pi \\ & & & 1 \end{bmatrix}, \begin{bmatrix} v_0' \\ v_1' \\ \tau' \\ 1 \end{bmatrix} = \mathscr{A}(F_p) \cdot \begin{bmatrix} v_0 \\ v_1 \\ \tau \\ 1 \end{bmatrix}$$

as a nontrivial isometric transformation leads to a group $\mathscr{A}\{\overline{\mathscr{F}}\}$ of order 16

$$\mathscr{A}\{\overline{\mathscr{F}}\} \overset{\text{is}}{=} \overline{\mathscr{F}}(\tau, v_0, v_1) \overset{\text{is}}{=} \mathscr{G}_{16}$$

This substitution group is contained in the direct sum $\Gamma\{\overline{\mathscr{F}}\}$ listed in Table 11b. $\mathscr{G}_{16}$ is identical with the full isometric group $\overline{\mathscr{H}}$ of this SRM since $\mathscr{G}(\tau, v_0, v_1) = C_1$, i.e. a NC with arbitrary values of the internal coordinates has no covering symmetry. The group $\mathscr{G}_{16}$ has the following structure:

(a) generators:

C and U,

$C^4 = U^2 = E,\ C^2U = UC^2$

$T = (CU)^2$ commutes with both C and U;

(b) subgroups of index 2 [k = n (mod 4)]:

$\{C^k, TC^k\} \overset{\text{is}}{=} C_{4h},$

$\{E, T, CU, CUT\} \times \{E, C^2\} \overset{\text{is}}{=} C_{4h}$

$\{E, C^2, C^2T, C^2U\} \times \{E, T\} \overset{\text{is}}{=} \mathscr{V}_4 \times \mathscr{V}_2 \overset{\text{is}}{=} \mathscr{A}(2, 2, 2);$

(c) center of $\mathscr{G}_{16}$

$\{E, T, C^2, TC^2\} \overset{\text{is}}{=} \mathscr{V}_4;$

(d) commutator group:

$[\mathscr{G}_{16}] = \{E, TC^2\};$

(e) important homomorphisms:

$\mathscr{G}_{16}/\{E, T\} \overset{\text{is}}{=} \mathscr{V}_4$

$\mathscr{G}_{16}/\{E, C^2\} \overset{\text{is}}{=} \mathscr{V}_4$

$\mathscr{G}_{16}/\{E, TC^2\} \overset{\text{is}}{=} C_{4h} \overset{\text{is}}{=} \mathscr{C}_4 \otimes \mathscr{V}_2$

Since $\mathscr{G}_{16}$ is homomorphic to ϑ_4 but does not contain $\mathscr{V}_4$ as a subgroup, we have

$$\overline{\mathscr{F}} \overset{\text{ho}}{=} \mathscr{F}, \mathscr{F} \not\subset \overline{\mathscr{F}}$$

As a next step the representation $\Gamma^{(NCf)}\{\overline{\mathscr{F}}\}$ induced by the substitutions $\xi' = F(\xi)$, $F \in \overline{\mathscr{F}}$ on the basis

$$\{\widetilde{X}_{t0}(v_0, \tau)\widetilde{X}_{t1}(v_1, \tau)\widetilde{X}_{f00}(\tau)\widetilde{X}_{f01}(\tau)\widetilde{X}_{f10}(\tau)\widetilde{X}_{f11}(\tau)\}$$

has to be calculated. Two sets of equivalent nuclei have to be considered since the coordinate vectors of the equivalent top nuclei do not generate a faithful representation of $\overline{\mathscr{F}}(\tau, v_0, v_1)$. Application of the operators $\hat{P}_F$ associated with the substitutions $\mathscr{A}(F)$ generates $\Gamma^{(NCf)}\{\overline{\mathscr{F}}\}$ listed in Table 11b

$$\Gamma^{(NCf)}\{\overline{\mathscr{F}}\} \overset{\text{is}}{=} \mathscr{G}_{16}$$

The rotational parts $\Gamma^{(3)}(F)$ of $\Gamma^{(NCf)}\{\overline{\mathscr{F}}\}$ form the group

$$\Gamma^{(3)}\{\overline{\mathscr{K}}\} = C_{4h}$$
$$\Delta^{(3)}\{\overline{\mathscr{K}}\} = C_4$$

From the last representation we get

$$\mathscr{B}\{\overline{\mathscr{F}}\} \overset{\text{is}}{=} C_4$$

and $\Gamma\{\overline{\mathscr{F}}\} \overset{\text{is}}{=} \mathscr{G}_{16}$, cf. Table 11b.

As may be derived from Table 11, the isometric substitution group $\overline{\mathscr{F}}$ has fixed points:

U: $v_0 = -v_1, \tau = 0$ $\quad\quad$ TC^2U: $v_0 = \pi - v_1, \tau = \pi/2$,

NCs with covering symmetry S^f_{12};

C^2U: $v_0 = \pi - v_1, \tau = 0$ $\quad$ TU: $v_0 = -v_1, \tau = \pi/2$,

NCs with covering symmetry Z.

The Longuet-Higgins group $\mathscr{L}\mathscr{H}$ is homomorphic to $\overline{\mathscr{F}}(\Gamma^{(NCl)}(F_p) = 1^{(6)} \otimes 1^{(3)})$

$$\mathscr{L}\mathscr{H} \overset{\text{is}}{=} \Gamma^{(NCl)}\{\overline{\mathscr{F}}\} \overset{\text{is}}{=} \mathscr{V}_4$$
$$\mathscr{L}\mathscr{H} \overset{\text{ho}}{=} \overline{\mathscr{F}}(\tau, v_0, v_1)$$

Table 12. Isometric group of the $D_2(\tau)F(\underline{C_1}TR)_2$ system
$\mathscr{G}(\tau, v_0, v_1) \overset{\text{is}}{=} \mathscr{C}_1$, $\mathscr{F}(\tau, v_0, v_1) \overset{\text{is}}{=} \vartheta_4$, $\overline{\mathscr{F}}(\tau, v_0, v_1) \overset{\text{is}}{=} \mathscr{G}_{16}$

Operator $F \in \mathscr{F}$	$\mathscr{A}\{\mathscr{F}\}$[a]	$\Gamma^{(\mathscr{N}\mathscr{C})}\{\mathscr{F}\}$[b]	Operator $F \in \overline{\mathscr{F}}$	$\Gamma^{(NCf)}\{\overline{\mathscr{F}}\}$[c]
E	$\begin{bmatrix} 1 & \cdot & \cdot & \cdot \\ \cdot & 1 & \cdot & \cdot \\ \cdot & \cdot & 1 & \cdot \\ \cdot & \cdot & \cdot & 1 \end{bmatrix}$	$\begin{bmatrix} 1 & \cdot & \cdot & \cdot \\ \cdot & 1 & \cdot & \cdot \\ \cdot & \cdot & 1 & \cdot \\ \cdot & \cdot & \cdot & 1 \end{bmatrix}$	E, T	$\begin{bmatrix} 1 & \cdot & \cdot & \cdot & \cdot & \cdot \\ \cdot & 1 & \cdot & \cdot & \cdot & \cdot \\ \cdot & \cdot & 1 & \cdot & \cdot & \cdot \\ \cdot & \cdot & \cdot & 1 & \cdot & \cdot \\ \cdot & \cdot & \cdot & \cdot & 1 & \cdot \\ \cdot & \cdot & \cdot & \cdot & \cdot & 1 \end{bmatrix} \otimes \begin{bmatrix} 1 & \cdot & \cdot \\ \cdot & 1 & \cdot \\ \cdot & \cdot & 1 \end{bmatrix}, \begin{bmatrix} 1 & \cdot & \cdot & \cdot & \cdot & \cdot \\ \cdot & 1 & \cdot & \cdot & \cdot & \cdot \\ \cdot & \cdot & 1 & \cdot & \cdot & \cdot \\ \cdot & \cdot & \cdot & 1 & \cdot & \cdot \\ \cdot & \cdot & \cdot & \cdot & 1 & \cdot \\ \cdot & \cdot & \cdot & \cdot & \cdot & 1 \end{bmatrix} \otimes \begin{bmatrix} -1 & \cdot & \cdot \\ \cdot & -1 & \cdot \\ \cdot & \cdot & 1 \end{bmatrix}$
F_2	$\begin{bmatrix} \cdot & 1 & \cdot & \cdot \\ 1 & \cdot & \cdot & \pi \\ \cdot & \cdot & 1 & \pi/2 \\ \cdot & \cdot & \cdot & 1 \end{bmatrix}$	$\begin{bmatrix} \cdot & \cdot & \cdot & 1 \\ \cdot & \cdot & 1 & \cdot \\ 1 & \cdot & \cdot & \cdot \\ \cdot & 1 & \cdot & \cdot \end{bmatrix}$	C, TC	$\begin{bmatrix} \cdot & 1 & \cdot & \cdot & \cdot & \cdot \\ 1 & \cdot & \cdot & \cdot & \cdot & \cdot \\ \cdot & \cdot & \cdot & \cdot & 1 & \cdot \\ \cdot & \cdot & \cdot & \cdot & \cdot & 1 \\ \cdot & \cdot & \cdot & 1 & \cdot & \cdot \\ \cdot & \cdot & 1 & \cdot & \cdot & \cdot \end{bmatrix} \otimes \begin{bmatrix} \cdot & -1 & \cdot \\ -1 & \cdot & \cdot \\ \cdot & \cdot & -1 \end{bmatrix}, \begin{bmatrix} \cdot & 1 & \cdot & \cdot & \cdot & \cdot \\ 1 & \cdot & \cdot & \cdot & \cdot & \cdot \\ \cdot & \cdot & \cdot & \cdot & 1 & \cdot \\ \cdot & \cdot & \cdot & \cdot & \cdot & 1 \\ \cdot & \cdot & \cdot & 1 & \cdot & \cdot \\ \cdot & \cdot & 1 & \cdot & \cdot & \cdot \end{bmatrix} \otimes \begin{bmatrix} \cdot & 1 & \cdot \\ 1 & \cdot & \cdot \\ \cdot & \cdot & -1 \end{bmatrix}$
F_3	$\begin{bmatrix} 1 & \cdot & \cdot & \pi \\ \cdot & 1 & \cdot & \pi \\ \cdot & \cdot & 1 & \cdot \\ \cdot & \cdot & \cdot & 1 \end{bmatrix}$	$\begin{bmatrix} \cdot & 1 & \cdot & \cdot \\ 1 & \cdot & \cdot & \cdot \\ \cdot & \cdot & \cdot & 1 \\ \cdot & \cdot & 1 & \cdot \end{bmatrix}$	C^2, TC^2	$\begin{bmatrix} 1 & \cdot & \cdot & \cdot & \cdot & \cdot \\ \cdot & 1 & \cdot & \cdot & \cdot & \cdot \\ \cdot & \cdot & \cdot & 1 & \cdot & \cdot \\ \cdot & \cdot & 1 & \cdot & \cdot & \cdot \\ \cdot & \cdot & \cdot & \cdot & \cdot & 1 \\ \cdot & \cdot & \cdot & \cdot & 1 & \cdot \end{bmatrix} \otimes \begin{bmatrix} 1 & \cdot & \cdot \\ \cdot & 1 & \cdot \\ \cdot & \cdot & 1 \end{bmatrix}, \begin{bmatrix} 1 & \cdot & \cdot & \cdot & \cdot & \cdot \\ \cdot & 1 & \cdot & \cdot & \cdot & \cdot \\ \cdot & \cdot & \cdot & 1 & \cdot & \cdot \\ \cdot & \cdot & 1 & \cdot & \cdot & \cdot \\ \cdot & \cdot & \cdot & \cdot & \cdot & 1 \\ \cdot & \cdot & \cdot & \cdot & 1 & \cdot \end{bmatrix} \otimes \begin{bmatrix} -1 & \cdot & \cdot \\ \cdot & -1 & \cdot \\ \cdot & \cdot & 1 \end{bmatrix}$
F_4	$\begin{bmatrix} \cdot & 1 & \cdot & \pi \\ 1 & \cdot & \cdot & \cdot \\ \cdot & \cdot & 1 & \pi/2 \\ \cdot & \cdot & \cdot & 1 \end{bmatrix}$	$\begin{bmatrix} \cdot & \cdot & 1 & \cdot \\ \cdot & \cdot & \cdot & 1 \\ \cdot & 1 & \cdot & \cdot \\ 1 & \cdot & \cdot & \cdot \end{bmatrix}$	C^3, TC^3	$\begin{bmatrix} \cdot & 1 & \cdot & \cdot & \cdot & \cdot \\ 1 & \cdot & \cdot & \cdot & \cdot & \cdot \\ \cdot & \cdot & \cdot & \cdot & \cdot & 1 \\ \cdot & \cdot & \cdot & \cdot & 1 & \cdot \\ \cdot & \cdot & 1 & \cdot & \cdot & \cdot \\ \cdot & \cdot & \cdot & 1 & \cdot & \cdot \end{bmatrix} \otimes \begin{bmatrix} \cdot & -1 & \cdot \\ -1 & \cdot & \cdot \\ \cdot & \cdot & -1 \end{bmatrix}, \begin{bmatrix} \cdot & 1 & \cdot & \cdot & \cdot & \cdot \\ 1 & \cdot & \cdot & \cdot & \cdot & \cdot \\ \cdot & \cdot & \cdot & \cdot & \cdot & 1 \\ \cdot & \cdot & \cdot & \cdot & 1 & \cdot \\ \cdot & \cdot & 1 & \cdot & \cdot & \cdot \\ \cdot & \cdot & \cdot & 1 & \cdot & \cdot \end{bmatrix} \otimes \begin{bmatrix} \cdot & 1 & \cdot \\ 1 & \cdot & \cdot \\ \cdot & \cdot & -1 \end{bmatrix}$

F_5	$\begin{bmatrix} \cdot & 1 & \cdot & \cdot \\ 1 & \cdot & \cdot & \cdot \\ \cdot & \cdot & 1 & \cdot \\ \cdot & \cdot & \cdot & 1 \end{bmatrix}$	$\begin{bmatrix} \cdot & \cdot & 1 & \cdot \\ \cdot & \cdot & \cdot & 1 \\ 1 & \cdot & \cdot & \cdot \\ \cdot & 1 & \cdot & \cdot \end{bmatrix}$	U, TU	$\begin{bmatrix} \cdot & 1 & \cdot & \cdot & \cdot & \cdot \\ 1 & \cdot & \cdot & \cdot & \cdot & \cdot \\ \cdot & \cdot & \cdot & \cdot & 1 & \cdot \\ \cdot & \cdot & \cdot & \cdot & \cdot & 1 \\ \cdot & \cdot & 1 & \cdot & \cdot & \cdot \\ \cdot & \cdot & \cdot & 1 & \cdot & \cdot \end{bmatrix} \otimes \begin{bmatrix} 1 & \cdot & \cdot \\ \cdot & -1 & \cdot \\ \cdot & \cdot & -1 \end{bmatrix}, \begin{bmatrix} \cdot & 1 & \cdot & \cdot & \cdot & \cdot \\ 1 & \cdot & \cdot & \cdot & \cdot & \cdot \\ \cdot & \cdot & \cdot & \cdot & 1 & \cdot \\ \cdot & \cdot & \cdot & \cdot & \cdot & 1 \\ \cdot & \cdot & 1 & \cdot & \cdot & \cdot \\ \cdot & \cdot & \cdot & 1 & \cdot & \cdot \end{bmatrix} \otimes \begin{bmatrix} -1 & \cdot & \cdot \\ \cdot & 1 & \cdot \\ \cdot & \cdot & -1 \end{bmatrix}$
F_6	$\begin{bmatrix} 1 & \cdot & \cdot & \cdot \\ \cdot & 1 & \cdot & \pi \\ \cdot & \cdot & 1 & \pi/2 \\ \cdot & \cdot & \cdot & 1 \end{bmatrix}$	$\begin{bmatrix} \cdot & 1 & \cdot & \cdot \\ 1 & \cdot & \cdot & \cdot \\ \cdot & \cdot & 1 & \cdot \\ \cdot & \cdot & \cdot & 1 \end{bmatrix}$	CU, TCU	$\begin{bmatrix} 1 & \cdot & \cdot & \cdot & \cdot & \cdot \\ \cdot & 1 & \cdot & \cdot & \cdot & \cdot \\ \cdot & \cdot & 1 & \cdot & \cdot & \cdot \\ \cdot & \cdot & \cdot & 1 & \cdot & \cdot \\ \cdot & \cdot & \cdot & \cdot & \cdot & 1 \\ \cdot & \cdot & \cdot & \cdot & 1 & \cdot \end{bmatrix} \otimes \begin{bmatrix} \cdot & 1 & \cdot \\ -1 & \cdot & \cdot \\ \cdot & \cdot & 1 \end{bmatrix}, \begin{bmatrix} 1 & \cdot & \cdot & \cdot & \cdot & \cdot \\ \cdot & 1 & \cdot & \cdot & \cdot & \cdot \\ \cdot & \cdot & 1 & \cdot & \cdot & \cdot \\ \cdot & \cdot & \cdot & 1 & \cdot & \cdot \\ \cdot & \cdot & \cdot & \cdot & \cdot & 1 \\ \cdot & \cdot & \cdot & \cdot & 1 & \cdot \end{bmatrix} \otimes \begin{bmatrix} \cdot & -1 & \cdot \\ 1 & \cdot & \cdot \\ \cdot & \cdot & 1 \end{bmatrix}$
F_7	$\begin{bmatrix} \cdot & 1 & \cdot & \pi \\ 1 & \cdot & \cdot & \pi \\ \cdot & \cdot & 1 & \cdot \\ \cdot & \cdot & \cdot & 1 \end{bmatrix}$	$\begin{bmatrix} \cdot & \cdot & \cdot & 1 \\ \cdot & \cdot & 1 & \cdot \\ \cdot & 1 & \cdot & \cdot \\ 1 & \cdot & \cdot & \cdot \end{bmatrix}$	C^2U, TC^2U	$\begin{bmatrix} \cdot & 1 & \cdot & \cdot & \cdot & \cdot \\ 1 & \cdot & \cdot & \cdot & \cdot & \cdot \\ \cdot & \cdot & \cdot & \cdot & \cdot & 1 \\ \cdot & \cdot & \cdot & \cdot & 1 & \cdot \\ \cdot & \cdot & \cdot & 1 & \cdot & \cdot \\ \cdot & \cdot & 1 & \cdot & \cdot & \cdot \end{bmatrix} \otimes \begin{bmatrix} 1 & \cdot & \cdot \\ \cdot & -1 & \cdot \\ \cdot & \cdot & -1 \end{bmatrix} \quad \begin{bmatrix} \cdot & 1 & \cdot & \cdot & \cdot & \cdot \\ 1 & \cdot & \cdot & \cdot & \cdot & \cdot \\ \cdot & \cdot & \cdot & \cdot & \cdot & 1 \\ \cdot & \cdot & \cdot & \cdot & 1 & \cdot \\ \cdot & \cdot & \cdot & 1 & \cdot & \cdot \\ \cdot & \cdot & 1 & \cdot & \cdot & \cdot \end{bmatrix} \otimes \begin{bmatrix} -1 & \cdot & \cdot \\ \cdot & 1 & \cdot \\ \cdot & \cdot & -1 \end{bmatrix}$
F_8	$\begin{bmatrix} 1 & \cdot & \cdot & \pi \\ \cdot & 1 & \cdot & \cdot \\ \cdot & \cdot & 1 & \pi/2 \\ \cdot & \cdot & \cdot & 1 \end{bmatrix}$	$\begin{bmatrix} 1 & \cdot & \cdot & \cdot \\ \cdot & 1 & \cdot & \cdot \\ \cdot & \cdot & \cdot & 1 \\ \cdot & \cdot & 1 & \cdot \end{bmatrix}$	C^3U, TC^3U	$\begin{bmatrix} 1 & \cdot & \cdot & \cdot & \cdot & \cdot \\ \cdot & 1 & \cdot & \cdot & \cdot & \cdot \\ \cdot & \cdot & \cdot & 1 & \cdot & \cdot \\ \cdot & \cdot & 1 & \cdot & \cdot & \cdot \\ \cdot & \cdot & \cdot & \cdot & 1 & \cdot \\ \cdot & \cdot & \cdot & \cdot & \cdot & 1 \end{bmatrix} \otimes \begin{bmatrix} \cdot & 1 & \cdot \\ -1 & \cdot & \cdot \\ \cdot & \cdot & 1 \end{bmatrix}, \begin{bmatrix} 1 & \cdot & \cdot & \cdot & \cdot & \cdot \\ \cdot & 1 & \cdot & \cdot & \cdot & \cdot \\ \cdot & \cdot & \cdot & 1 & \cdot & \cdot \\ \cdot & \cdot & 1 & \cdot & \cdot & \cdot \\ \cdot & \cdot & \cdot & \cdot & 1 & \cdot \\ \cdot & \cdot & \cdot & \cdot & \cdot & 1 \end{bmatrix} \otimes \begin{bmatrix} \cdot & -1 & \cdot \\ 1 & \cdot & \cdot \\ \cdot & \cdot & 1 \end{bmatrix}$

[a] Representation of $\mathscr{F}$ by substitutions of the internal coordinates: $\begin{bmatrix} v_0' \\ v_1' \\ \tau' \\ 1 \end{bmatrix} = \begin{bmatrix} A(F) & a(F) \\ 0 & 1 \end{bmatrix} \begin{bmatrix} v_0 \\ v_1 \\ \tau \\ 1 \end{bmatrix}$ $\quad \mathscr{F}\colon -\frac{\pi}{2} < \tau \leqslant +\frac{\pi}{2}$, τ mod p

$\overline{\mathscr{F}}\colon -\pi < \tau \leqslant \pi$, $\quad \tau$ mod 2p

$-\pi < v_0, v_1 \leqslant \pi$

[b] Representation by permutations generated by the set of distances $d_{t0,\,f10}(v_0, \tau)$, $d_{t0,\,f11}(v_0, \tau)$, $d_{t1,\,f00}(v_1, \tau)$, $d_{t1,\,f01}(v_1, \tau)$ (see Fig. 5a).

[c] Representation generated by the vectors $\tilde{X}_{t0}(v_0, \tau)$, $\tilde{X}_{t1}(v_1, \tau)$, $\tilde{X}_{f00}(\tau)$, $\tilde{X}_{f01}(\tau)$, $\tilde{X}_{f10}(\tau)$, $\tilde{X}_{f11}(\tau)$ of two sets of equivalent nuclei originating from the two equivalent tops and the frame, respectively (see Fig. 5a).

(ii) $D_2(\tau)F(C_1TR)_2$ System

This case follows closely the treatment given for the $D_2(\tau)F(C_1TR)(C_1TS)$ system. The relevant groups are listed in Table 12

$$\mathcal{G}(\tau, v_0, v_1) = C_1$$
$$\mathcal{A}\{\mathcal{F}\} \overset{\text{is}}{=} \Gamma^{(\mathcal{N}\mathcal{C})}\{\mathcal{F}\} \overset{\text{is}}{=} \vartheta_4$$
$$\Gamma^{(\text{NCf})}\{\overline{\mathcal{F}}\} \overset{\text{is}}{=} \mathcal{G}_{16}$$

The abstract group $\mathcal{G}_{16}$ is identical with the group $\mathcal{G}_{16}$ of the $D_2(\tau)F(C_1TR)(C_1TS)$ model. In contrast to the (R, S) system

$$\Gamma^{(3)}\{\overline{\mathcal{K}}\} = D_4$$

(*case a,* cf. Sect. 2.2.1). Therefore, all fixed point NCs (only F_5 and F_7 possess fixed points) must have properly orthogonal covering operations ($C_2(e_1^f)$ and $C_2(e_2^f)$, respectively).

2.4.4 $D_{\infty h}F(C_{2v}T)_2$ System

Molecules with a frame of local symmetry $D_{\infty h}$ and two equivalent tops with local symmetry C_{2v}, e.g. biphenyl ($(C_6H_5)_2$), ethylene ($CH_2{=}CH_2$) or the bicyclic organoboron compound $(\overline{CH_2CH_2B})_2$, are examples for a SRM with proper covering symmetry $\mathcal{G}$ (ξ). A model of $(\overline{CH_2CH_2B})_2$ is illustrated in Fig. 6 together with the frame

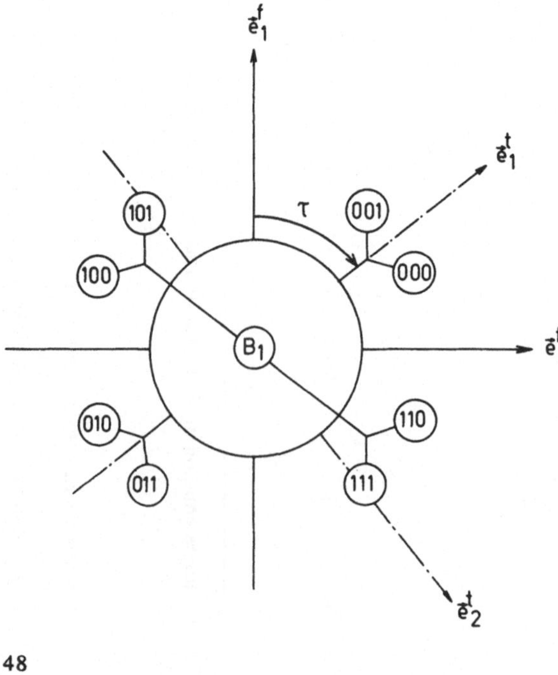

Fig. 6. Newman projection of $D_{\infty h}F(C_{2v}T)_2$ system $(\overline{CH_2CH_2B})_2$
Key: Only B and H nuclei are shown. The origin of the frame system $\tilde{e}^f$ is situated in the center of the B-B bond, the origin of the local top coordinate system $\tilde{e}^t$ lies in the nucleus B_0

coordinate system and the choice of the internal coordinate τ. Coordinate vectors of a set of eight equivalent top nuclei in general site are (w.r.t. $\tilde{\mathfrak{e}}^f$)

$$\widetilde{X}_{\lambda\mu\nu}(\tau) = \left\{ \frac{r}{2}(001) + \widetilde{X}^t_{000} \begin{bmatrix} (-1)^{\mu} & & \\ & (-1)^{\mu+\nu} & \\ & & 1 \end{bmatrix} \begin{bmatrix} c\tau & s\tau & 0 \\ -s\tau & c\tau & 0 \\ 0 & 0 & 1 \end{bmatrix} \right\} \begin{bmatrix} 1 & & \\ & -1 & \\ & & -1 \end{bmatrix}^{\lambda}$$

where r denotes the B–B bond length and $\widetilde{X}^t_{000} = (X^t_{0001} X^t_{0002} X^t_{0003})$ is the coordinate vector of the representative nucleus w.r.t. the local top coordinate system $\tilde{\mathfrak{e}}^t$ indicated in Fig. 6. As mentioned earlier it is important to take a set of nuclei in general site, otherwise the permutation groups $\Pi\{\mathscr{F}\}(\Pi\{\mathscr{H}\})$ and $\Gamma^{(\mathscr{N}\mathscr{C})}\{\mathscr{F}\}$ $(\Gamma^{(\mathscr{N}\mathscr{C})}\{\mathscr{H}\})$ will not be faithful representations of $\mathscr{F}(\tau)(\mathscr{H}(\tau))$.

From the distance formula

$$\begin{aligned} d^2_{0\mu\nu,\,1\bar{\mu}\bar{\nu}}(\tau) &= |X_{0\mu\nu}(\tau) - X_{1\bar{\mu}\bar{\nu}}(\tau)|^2 \\ &= r^2 + 4\, r X^t_{0003} + 2\, \widetilde{X}^t_{000} X^t_{000} + \\ &+ 2\, \widetilde{X}^t_{000} \begin{bmatrix} (-1)^{\bar{\mu}+\mu+1} c2\tau & (-1)^{\bar{\mu}+\mu+\nu} s2\tau & 0 \\ (-1)^{\bar{\mu}+\mu+\bar{\nu}} s2\tau & (-1)^{\bar{\mu}+\bar{\nu}+\mu+\nu} c2\tau & 0 \\ 0 & 0 & 1 \end{bmatrix} X^t_{000} \end{aligned}$$

$$\mu, \nu, \bar{\mu}, \bar{\nu} = 0,1$$

we conclude that the primitive period p of τ is equal to π. For the domain $-\pi/2 < \tau \leqslant +\pi/2$ we have the isometric substitution group

$$\mathscr{A}\{\mathscr{F}\} := \left\{ \begin{pmatrix} 1 & \cdot \\ \cdot & 1 \end{pmatrix}, \begin{pmatrix} -1 & \cdot \\ \cdot & 1 \end{pmatrix}, \begin{pmatrix} -1 & \pi/2 \\ \cdot & 1 \end{pmatrix}, \begin{pmatrix} 1 & \pi/2 \\ \cdot & 1 \end{pmatrix} \right\} \overset{\text{is}}{=} \mathscr{V}_4$$

$$\begin{pmatrix} \tau' \\ 1 \end{pmatrix} = \mathscr{A}(F) \begin{pmatrix} \tau \\ 1 \end{pmatrix}$$

$\tau' = \tau + \pi$ is a nontrivial primitive period transformation since the coordinate vectors $X_{\lambda\mu\nu}(\tau)$ are periodic in τ with period $2\,p = 2\,\pi$. $\mathscr{A}\{\mathscr{F}\}$ has therefore to be extended to $\mathscr{A}\{\overline{\mathscr{F}}\}$ by taking τ modulo $2\,\pi$

$$-\pi < \tau \leqslant +\pi$$

$$\overline{\mathscr{F}}(\tau) \overset{\text{is}}{=} \mathscr{A}\{\overline{\mathscr{F}}\} \overset{\text{is}}{=} \vartheta_4, \text{ cf. Table 13a}$$

A NC with arbitrary τ has covering symmetry D_2, hence

$$\mathscr{G}(\tau) \overset{\text{is}}{=} \Gamma^{(3)}\{\mathscr{G}\} := \left\{ \begin{pmatrix} 1 & & \\ & 1 & \\ & & 1 \end{pmatrix}, \begin{pmatrix} 1 & & \\ & -1 & \\ & & -1 \end{pmatrix}, \begin{pmatrix} -1 & & \\ & 1 & \\ & & -1 \end{pmatrix}, \begin{pmatrix} -1 & & \\ & -1 & \\ & & 1 \end{pmatrix} \right\} = D_2$$

cf. Fig. 6.

Table 13a. Internal isometric group of the $D_{\infty h}F(C_{2v}T)_2$ system
$\mathscr{F}(\tau) \overset{is}{\equiv} \vartheta_4$, $\mathscr{F}(\tau) \overset{is}{\equiv} \mathscr{V}_4$

Operator[a] $F \in \mathscr{F}$	$\mathscr{A}\{\bar{\mathscr{F}}\}$[b]	$\Gamma^{(NCf)}\{\bar{\mathscr{F}}\}$[c]	$\Gamma\{\bar{\mathscr{F}}\}$[d]
E	$\begin{bmatrix} 1 & \cdot \\ \cdot & 1 \end{bmatrix}$	$(\delta_{\lambda'\lambda}\delta_{\mu'\mu}\delta_{\nu'\nu}) \otimes \begin{bmatrix} 1 & \cdot & \cdot \\ & 1 & \cdot \\ & & 1 \end{bmatrix}$	$\begin{bmatrix} 1 & \cdot & \cdot & \cdot & \cdot \\ & 1 & \cdot & \cdot & \cdot \\ & & 1 & \cdot & \cdot \\ & & & 1 & \cdot \\ & & & & 1 \end{bmatrix}$
$F_2(SW)$	$\begin{bmatrix} -1 & \cdot \\ & 1 \end{bmatrix}$	$(\delta_{\lambda'\lambda+1}\delta_{\mu'\mu}\delta_{\nu'\nu+1}) \otimes \begin{bmatrix} 1 & \cdot & \cdot \\ & 1 & \cdot \\ & & -1 \end{bmatrix}$	$\begin{bmatrix} 1 & \cdot & \cdot & \cdot & \cdot \\ & 1 & \cdot & \cdot & \cdot \\ & & 1 & \cdot & \pi \\ & & & -1 & \cdot \\ & & & & 1 \end{bmatrix}$
$F_3(TC^3S)$	$\begin{bmatrix} -1 & \pi/2 \\ \cdot & 1 \end{bmatrix}$	$(\delta_{\lambda'\lambda}\delta_{\mu'\mu+\lambda}\delta_{\nu'\nu+1}) \otimes \begin{bmatrix} \cdot & 1 & \cdot \\ 1 & \cdot & \cdot \\ \cdot & \cdot & 1 \end{bmatrix}$	$\begin{bmatrix} 1 & \cdot & \cdot & \cdot & \pi \\ & -1 & \cdot & \cdot & \pi \\ & & -1 & \cdot & \pi/2 \\ & & & -1 & \pi/2 \\ & & & & 1 \end{bmatrix}$
$F_4(CW)$	$\begin{bmatrix} 1 & -\pi/2 \\ \cdot & 1 \end{bmatrix}$	$(\delta_{\lambda'\lambda+1}\delta_{\mu'\mu+\lambda}\delta_{\nu'\nu}) \otimes \begin{bmatrix} \cdot & 1 & \cdot \\ 1 & \cdot & \cdot \\ \cdot & \cdot & -1 \end{bmatrix}$	$\begin{bmatrix} 1 & \cdot & \cdot & \cdot & \pi \\ & -1 & \cdot & \cdot & \pi \\ & & -1 & \cdot & 3\pi/2 \\ & & & 1 & -\pi/2 \\ & & & & 1 \end{bmatrix}$
$F_5(TC^2) = F_p$	$\begin{bmatrix} 1 & \pi \\ \cdot & 1 \end{bmatrix}$	$(\delta_{\lambda'\lambda}\delta_{\mu'\mu+1}\delta_{\nu'\nu}) \otimes \begin{bmatrix} 1 & \cdot & \cdot \\ & 1 & \cdot \\ & & 1 \end{bmatrix}$	$\begin{bmatrix} 1 & \cdot & \cdot & \cdot & \cdot \\ & 1 & \cdot & \cdot & \cdot \\ & & 1 & \cdot & \cdot \\ & & & 1 & \pi \\ & & & & 1 \end{bmatrix}$

$F_6(TC^2SW)$	$\begin{bmatrix} -1 & \pi \\ \cdot & 1 \end{bmatrix}$	$({}^{\delta}\lambda'\lambda + 1{}^{\delta}\mu'\mu + 1{}^{\delta}\nu'\nu + 1) \otimes \begin{bmatrix} 1 & \cdot & \cdot \\ & 1 & \cdot \\ & & -1 \end{bmatrix}$	$\begin{bmatrix} 1 & \cdot & \cdot & \cdot & \cdot \\ & 1 & \cdot & \cdot & \cdot \\ & & 1 & \cdot & \pi \\ & & & -1 & \pi \\ & & & & 1 \end{bmatrix}$
$F_7(CS)$	$\begin{bmatrix} -1 & -\pi/2 \\ \cdot & 1 \end{bmatrix}$	$({}^{\delta}\lambda'\lambda{}^{\delta}\mu'\mu + \lambda + 1{}^{\delta}\nu'\nu + 1) \otimes \begin{bmatrix} \cdot & 1 & \cdot \\ 1 & \cdot & \cdot \\ \cdot & \cdot & 1 \end{bmatrix}$	$\begin{bmatrix} 1 & \cdot & \cdot & \cdot & \pi \\ & -1 & \cdot & \cdot & \pi \\ & & -1 & \cdot & \pi/2 \\ & & & -1 & -\pi/2 \\ & & & & 1 \end{bmatrix}$
$F_8(TC^3W)$	$\begin{bmatrix} 1 & \pi/2 \\ \cdot & 1 \end{bmatrix}$	$({}^{\delta}\lambda'\lambda + 1{}^{\delta}\mu'\mu + \lambda + 1{}^{\delta}\nu'\nu) \otimes \begin{bmatrix} \cdot & 1 & \cdot \\ 1 & \cdot & \cdot \\ \cdot & \cdot & -1 \end{bmatrix}$	$\begin{bmatrix} 1 & \cdot & \cdot & \cdot & \pi \\ & -1 & \cdot & \cdot & \pi \\ & & -1 & \cdot & 3\pi/2 \\ & & & 1 & \pi/2 \\ & & & & 1 \end{bmatrix}$

a Symbols in parentheses denote elements of the abstract group $\mathcal{G}_{32}$.

b Representation of $\mathcal{F}$ by substitutions of the internal coordinate:

$$\begin{bmatrix} \tau' \\ 1 \end{bmatrix} = \begin{bmatrix} A(F) & a(F) \\ 0 & 1 \end{bmatrix} \cdot \begin{bmatrix} \tau \\ 1 \end{bmatrix}, \quad -\pi < \tau \leqslant +\pi, \ \tau \bmod 2\,p.$$

c Representation generated by the vectors $\widetilde{X}_{000}(\tau)$, $\widetilde{X}_{001}(\tau)$, $\widetilde{X}_{010}(\tau)$, $\widetilde{X}_{011}(\tau)$, $\widetilde{X}_{100}(\tau)$, $\widetilde{X}_{101}(\tau)$, $\widetilde{X}_{110}(\tau)$, $\widetilde{X}_{111}(\tau)$ of a set of equivalent nuclei from the two tops (see Fig. 6). For the notation of the permutation matrices Π by Kronecker symbols see Appendix 3.

d Representation by substitutions of the eulerian angles and internal coordinates:

$$\begin{bmatrix} \alpha' \\ \beta' \\ \gamma' \\ \tau' \\ 1 \end{bmatrix} = \begin{bmatrix} B(F) & 0 & b(F) \\ \cdot & A(F) & a(F) \\ \cdot & \cdot & 1 \end{bmatrix} \cdot \begin{bmatrix} \alpha \\ \beta \\ \gamma \\ \tau \\ 1 \end{bmatrix}$$

The representation of the full isometric group $\mathscr{H}(\tau)$ on the nuclear position vectors, $\Gamma^{(NCf)}\{\mathscr{H}\}$, may then be calculated by solving Eq. (2.69), Sect. 2.2.4. for all

$$F \in \mathscr{F}(\tau).$$

$$\hat{P}_F\{\tilde{X}_{\lambda\mu\nu}(\tau)\} = \{\tilde{X}_{\lambda\mu\nu}(F^{-1}(\tau))\} = \{\tilde{X}_{\lambda\mu\nu}(\tau)\}\Pi(F) \otimes \Gamma^{(3)}(F)$$

or for one vector

$$\hat{P}_F\tilde{X}_{\lambda\mu\nu}(\tau) = \tilde{X}_{\lambda\mu\nu}(F^{-1}(\tau)) = \tilde{X}_{\lambda'\mu'\nu'}\Gamma^{(3)}(F) \qquad (2.69')$$

Each of these equations admits $|\mathscr{G}| = 4$ solutions. An example for the solution of this equation is given in Appendix 3. In particular, the invariant subgroup $\Gamma^{(NCf)}\{\mathscr{G}\}$ of $\Gamma^{(NCf)}\{\mathscr{H}\}$ is obtained by solving the equation

$$\tilde{X}_{\lambda\mu\nu}(\tau) = \tilde{X}_{\lambda'\mu'\nu'}(\tau)\Gamma^{(3)}(G), \forall G \in \mathscr{G}$$

Table 13b. Covering group of the $D_{\infty h}F(C_{2v}T)_2$ system
$\mathscr{G}(\tau) = D_2$

Operator[a] $G \in \mathscr{G}$	$\Gamma^{(NCf)}\{\mathscr{G}\}^c$	$\Gamma\{\mathscr{G}\}^d$
E	$(\delta_{\lambda'\lambda}\delta_{\mu'\mu}\delta_{\nu'\nu}) \otimes \begin{bmatrix} 1 & \cdot & \cdot \\ & 1 & \cdot \\ & & 1 \end{bmatrix}$	$\begin{bmatrix} 1 & \cdot & \cdot & \cdot & \cdot \\ & 1 & \cdot & \cdot & \cdot \\ & & 1 & \cdot & \cdot \\ & & & 1 & \cdot \\ & & & & 1 \end{bmatrix}$
$G_2(W)$	$(\delta_{\lambda'\lambda+1}\delta_{\mu'\mu}\delta_{\nu'\nu}) \otimes \begin{bmatrix} 1 & \cdot & \cdot \\ & -1 & \cdot \\ & & -1 \end{bmatrix}$	$\begin{bmatrix} 1 & \cdot & \cdot & \cdot & \pi \\ & -1 & \cdot & \cdot & \pi \\ & & -1 & \cdot & \cdot \\ & & & 1 & \cdot \\ & & & & 1 \end{bmatrix}$
$G_3(TW)$	$(\delta_{\lambda'\lambda+1}\delta_{\mu'\mu+1}\delta_{\nu'\nu}) \otimes \begin{bmatrix} -1 & \cdot & \cdot \\ & 1 & \cdot \\ & & -1 \end{bmatrix}$	$\begin{bmatrix} 1 & \cdot & \cdot & \cdot & \pi \\ & -1 & \cdot & \cdot & \pi \\ & & -1 & \cdot & \pi \\ & & & 1 & \cdot \\ & & & & 1 \end{bmatrix}$
$G_4(T)$	$(\delta_{\lambda'\lambda}\delta_{\mu'\mu+1}\delta_{\nu'\nu}) \otimes \begin{bmatrix} -1 & \cdot & \cdot \\ & -1 & \cdot \\ & & 1 \end{bmatrix}$	$\begin{bmatrix} 1 & \cdot & \cdot & \cdot & \cdot \\ & 1 & \cdot & \cdot & \cdot \\ & & 1 & \cdot & \pi \\ & & & 1 & \cdot \\ & & & & 1 \end{bmatrix}$

[a, c, d] See caption of Table 13a.

Table 13c. Generators of the full isometric group $\overline{\mathscr{H}}(\tau)$ of the SRM $D_{\infty h}F(C_{2v}T)_2$
$\overline{\mathscr{H}}(\tau) \overset{\text{is}}{=} \mathscr{G}_{32}$

Generator[a]	$\Gamma^{(NCf)}\{\overline{\mathscr{H}}\}$[c]	$\Gamma\{\overline{\mathscr{H}}\}$[d]
C	$(\delta_{\lambda'\lambda}\delta_{\mu'\mu+\lambda+1}\delta_{\nu'\nu}) \otimes \begin{bmatrix} \cdot & -1 & \cdot \\ 1 & \cdot & \cdot \\ \cdot & \cdot & 1 \end{bmatrix}$	$\begin{bmatrix} 1 & \cdot & \cdot & \cdot & \cdot \\ & 1 & \cdot & \cdot & \cdot \\ & & 1 & \cdot & 3\pi/2 \\ & & \cdot & 1 & -\pi/2 \\ & & & & 1 \end{bmatrix}$
S	$(\delta_{\lambda'\lambda}\delta_{\mu'\mu}\delta_{\nu'\nu+1}) \otimes \begin{bmatrix} 1 & \cdot & \cdot \\ & -1 & \cdot \\ & & 1 \end{bmatrix}$	$\begin{bmatrix} 1 & \cdot & \cdot & \cdot & \pi \\ & -1 & \cdot & \cdot & \pi \\ & & -1 & \cdot & \pi \\ & & & -1 & \cdot \\ & & & & 1 \end{bmatrix}$
W	$(\delta_{\lambda'\lambda+1}\delta_{\mu'\mu}\delta_{\nu'\nu}) \otimes \begin{bmatrix} 1 & \cdot & \cdot \\ & -1 & \cdot \\ & & -1 \end{bmatrix}$	$\begin{bmatrix} 1 & \cdot & \cdot & \cdot & \pi \\ & -1 & \cdot & \cdot & \pi \\ & & -1 & \cdot & \cdot \\ & & & 1 & \cdot \\ & & & & 1 \end{bmatrix}$
$T = WCWC^3$	$(\delta_{\lambda'\lambda}\delta_{\mu'\mu+1}\delta_{\nu'\nu}) \otimes \begin{bmatrix} -1 & & \\ & -1 & \\ & & 1 \end{bmatrix}$	$\begin{bmatrix} 1 & \cdot & \cdot & \cdot & \cdot \\ & 1 & \cdot & \cdot & \cdot \\ & & 1 & \cdot & \pi \\ & & & 1 & \cdot \\ & & & & 1 \end{bmatrix}$

[a, c, d] See caption of Table 13a.

with the result

$$\Gamma^{(NCf)}\{\mathscr{G}\} \overset{\text{is}}{=} \mathscr{G}(\tau) \overset{\text{is}}{=} D_2 \overset{\text{is}}{=} \vartheta_2$$

(D_2 denoting the point symmetry group, ϑ_2 the abstract group). This representation is listed in Table 13b. The easiest way to get $\Gamma^{(NCf)}\{\overline{\mathscr{H}}\}$ is to calculate first $\Gamma^{(NCf)}\{\mathscr{G}\}$ and one solution $\Gamma^{(NCf)}(F_k)$ of Eq. (2.69′) for each $F \in \overline{\mathscr{F}}$. The set of all $|\mathscr{G}|$ solutions for each F is then equal to the coset

$$\Gamma^{(NCf)}(F_k) \cdot \Gamma^{(NCf)}\{\mathscr{G}\}$$

and $\Gamma^{(NCf)}\{\overline{\mathscr{H}}\} = \bigcup_{k=1}^{|\mathscr{F}|} \Gamma^{(NCf)}(F_k)\,\Gamma^{(NCf)}\{\mathscr{G}\}$

The full isometric group $\overline{\mathscr{H}}(\tau)$ in our example is isomorphic to a group of order $|\overline{\mathscr{F}}|\,|\mathscr{G}| = 32$

$$\overline{\mathscr{H}}(\tau) \overset{\text{is}}{=} \Gamma^{(NCf)}\{\overline{\mathscr{H}}\} \overset{\text{is}}{=} \mathscr{G}_{32}$$

The abstract group $\mathcal{G}_{32}$ has generators C, S and W with the relations

$$C^4 = S^2 = W^2 = E$$
$$SC^k = C^{-k}S, k = n \pmod 4$$
$$WS = SW$$

The element $T = WCWC^3$ commutes with C, S and W and

$$WC^k = T^kC^kW, T^2 = E$$

The generators of the representation $\Gamma^{(NCf)}\{\overline{\mathscr{H}}\}$ together with the element $\Gamma^{(NCf)}(T)$ are given in Table 13c. C, S and T are generators of a subgroup of $\mathcal{G}_{32}$ of index 2 isomorphic to D_{4h}

$$\vartheta_4[E, T] = [\{C^k\} \cup S\{C^k\}][E, T]$$
$$\mathcal{G}_{32} = \vartheta_4 \cdot [E, T] \cup W \cdot \vartheta_4 \cdot [E, T]$$

The representation $\Gamma^{(NCf)}\{\overline{\mathscr{F}}\}$ of the internal isometric group $\overline{\mathscr{F}}(\tau)$ is not uniquely determined since several groups $\Gamma^{(3)}\{\overline{\mathscr{K}}\}$ and, therefore, several inverse homomorphisms $\mathfrak{h}^{-1}(\Gamma^{(3)}\{\overline{\mathscr{K}}\}) = \Gamma^{(NCf)}\{\overline{\mathscr{F}}\}$ [Eq. (2.80)] may be defined, or in other words, several sets of representatives of the cosets of the invariant subgroup $\Gamma^{(NCf)}\{\mathcal{G}\}$ in $\Gamma^{(NCf)}\{\overline{\mathscr{H}}\}$ form a subgroup isomorphic to $\overline{\mathscr{F}}(\tau)$. One such representation is shown in Table 13a

$$\Gamma^{(NCf)}\{\overline{\mathscr{F}}\} \overset{is}{=} \overline{\mathscr{F}}(\tau) \overset{is}{=} \vartheta_4$$

Its semidirect product with $\Gamma^{(NCf)}\{\mathcal{G}\}$ gives $\Gamma^{(NCf)}\{\overline{\mathscr{H}}\}$

$$\Gamma^{(NCf)}\{\mathcal{G}\} \cdot \Gamma^{(NCf)}\{\overline{\mathscr{F}}\} = \vartheta_2 \cdot \vartheta_4 = \mathcal{G}_{32}$$

Various representations of $\overline{\mathscr{H}}(\tau)$ may be derived from $\Gamma^{(NCf)}\{\overline{\mathscr{H}}\}$.

(i) the set of all rotational parts $\Gamma^{(3)}(H)$ of $\Gamma^{(NCf)}\{\overline{\mathscr{H}}\}$ forms the point group D_{4h} (cf. Table 13)

$$\Gamma^{(3)}\{\overline{\mathscr{L}}\} = \Gamma^{(3)}\{\mathcal{G}\}\,\Gamma^{(3)}\{\overline{\mathscr{K}}\} = D_{4h}$$

$$\text{with } \Gamma^{(3)}\{\mathcal{G}\} := \left\{\begin{pmatrix}1&&\\&1&\\&&1\end{pmatrix}, \begin{pmatrix}1&&\\&-1&\\&&-1\end{pmatrix}, \begin{pmatrix}-1&&\\&1&\\&&-1\end{pmatrix}, \begin{pmatrix}-1&&\\&-1&\\&&1\end{pmatrix}\right\} = D_2$$

$$\Gamma^{(3)}\{\overline{\mathscr{K}}\} := \left\{\begin{pmatrix}1&&\\&1&\\&&1\end{pmatrix}, \begin{pmatrix}1&&\\&1&\\&&-1\end{pmatrix}, \begin{pmatrix}\cdot&1&\cdot\\1&\cdot&\cdot\\\cdot&\cdot&1\end{pmatrix}, \begin{pmatrix}\cdot&1&\cdot\\1&\cdot&\cdot\\\cdot&\cdot&-1\end{pmatrix}\right\} = C_{2v}$$

and $\Gamma^{(3)}\{\mathcal{G}\} \cap \Gamma^{(3)}\{\overline{\mathscr{K}}\} = 1^{(3)}$

(ii) the group of matrices $R(H) = |\Gamma^{(3)}(H)| \cdot \Gamma^{(3)}(H)$ is given by

$$\Delta^{(3)}\{\overline{\mathscr{L}}\} = \Delta^{(3)}\{\mathscr{G}\}\,\Delta^{(3)}\{\overline{\mathscr{K}}\} = D_4 \qquad \text{(case b 1)}$$

with

$$\Delta^{(3)}\{\mathscr{G}\} \equiv \Gamma^{(3)}\{\mathscr{G}\} = D_2$$

$$\Delta^{(3)}\{\overline{\mathscr{K}}\} := \left\{ \begin{pmatrix} 1 & & \\ & 1 & \\ & & 1 \end{pmatrix}, \begin{pmatrix} -1 & & \\ & -1 & \\ & & 1 \end{pmatrix}, \begin{pmatrix} \cdot & -1 & \cdot \\ -1 & \cdot & \cdot \\ \cdot & \cdot & -1 \end{pmatrix}, \begin{pmatrix} \cdot & 1 & \cdot \\ 1 & \cdot & \cdot \\ \cdot & \cdot & -1 \end{pmatrix} \right\} = D_2$$

$$\Delta^{(3)}\{\mathscr{G}\} \cap \Delta^{(3)}\{\overline{\mathscr{K}}\} = \left\{ \begin{pmatrix} 1 & & \\ & 1 & \\ & & 1 \end{pmatrix}, \begin{pmatrix} -1 & & \\ & -1 & \\ & & 1 \end{pmatrix} \right\}$$

(iii) from the representations $\Delta^{(3)}$ one gets the transformation groups of the eulerian angles

$$\mathscr{B}\{\mathscr{G}\} := \left\{ \begin{bmatrix} 1 & & & \\ & 1 & & \\ & & 1 & \\ & & & 1 \end{bmatrix}, \begin{bmatrix} 1 & \cdot & \cdot & \pi \\ & -1 & \cdot & \pi \\ & & -1 & \cdot \\ & & & 1 \end{bmatrix}, \begin{bmatrix} 1 & \cdot & \cdot & \pi \\ & -1 & \cdot & \pi \\ & & -1 & \pi \\ & & & 1 \end{bmatrix}, \begin{bmatrix} 1 & \cdot & \cdot & \cdot \\ & 1 & \cdot & \cdot \\ & & 1 & \pi \\ & & & 1 \end{bmatrix} \right\} \overset{\text{is}}{=} \vartheta_2$$

$$\mathscr{B}\{\overline{\mathscr{F}}\} := \left\{ \begin{bmatrix} 1 & & & \\ & 1 & & \\ & & 1 & \\ & & & 1 \end{bmatrix}, \begin{bmatrix} 1 & \cdot & \cdot & \cdot \\ & 1 & \cdot & \cdot \\ & & 1 & \pi \\ & & & 1 \end{bmatrix}, \begin{bmatrix} 1 & \cdot & \cdot & \pi \\ & -1 & \cdot & \pi \\ & & -1 & \pi/2 \\ & & & 1 \end{bmatrix}, \begin{bmatrix} 1 & \cdot & \cdot & \pi \\ & -1 & \cdot & \pi \\ & & -1 & \frac{3\pi}{2} \\ & & & 1 \end{bmatrix} \right\} \overset{\text{is}}{=} \vartheta_2$$

$$\mathscr{B}\{\overline{\mathscr{H}}\} = \mathscr{B}\{\mathscr{G}\} \cdot \mathscr{B}\{\overline{\mathscr{F}}\} \overset{\text{is}}{=} \vartheta_4$$

$$\mathscr{B}\{\mathscr{G}\} \cap \mathscr{B}\{\overline{\mathscr{F}}\} = \left\{ \begin{bmatrix} 1 & & & \\ & 1 & & \\ & & 1 & \\ & & & 1 \end{bmatrix}, \begin{bmatrix} 1 & \cdot & \cdot & \cdot \\ & 1 & \cdot & \cdot \\ & & 1 & \pi \\ & & & 1 \end{bmatrix} \right\}$$

(iv) and the substitution groups of the eulerian angles and internal coordinates

$$\Gamma\{\mathscr{G}\} := \{\mathscr{B}(G) \oplus 1^{(2)} \mid \forall G \in \mathscr{G}(\tau)\} \overset{\text{is}}{=} \vartheta_2$$
$$\Gamma\{\overline{\mathscr{F}}\} := \{\mathscr{B}(F) \oplus \mathscr{A}(F) \mid \forall F \in \overline{\mathscr{F}}(\tau)\} \overset{\text{is}}{=} \vartheta_4$$
$$\Gamma\{\overline{\mathscr{H}}\} := \Gamma\{\mathscr{G}\} \cdot \Gamma\{\overline{\mathscr{F}}\} = \vartheta_2 \cdot \vartheta_4 = \mathscr{G}_{32}$$
$$\Gamma\{\mathscr{G}\} \cap \Gamma\{\overline{\mathscr{F}}\} = 1$$

cf. Table 13.

The Longuet-Higgins group $\mathcal{L}\mathcal{H}$ is easily derived from $\Gamma^{(NCf)}\{\overline{\mathcal{H}}\}$

$$\Gamma^{(NCl)}\{\overline{\mathcal{H}}\} := \{\Pi(H) \otimes |\Gamma^{(3)}(H)| \cdot 1^{(3)} \mid \forall H \in \overline{\mathcal{H}}(\tau)\} \overset{is}{=} D_{4h}$$

Thus, $\mathcal{L}\mathcal{H}$ is only homomorphic to $\overline{\mathcal{H}}(\tau)$

$$\mathcal{L}\mathcal{H} \overset{is}{=} \Gamma^{(NCl)}\{\overline{\mathcal{H}}\} \overset{ho}{=} \overline{\mathcal{H}}(\tau) \overset{is}{=} \mathcal{G}_{32}$$

because one of the elements of the coset $\Gamma^{(NCf)}(F_p) \cdot \Gamma^{(NCf)}\{\mathcal{G}\}$ associated with the primitive period substitution $\tau' = \tau + \pi$ is equal to $1^{(8)} \otimes \begin{pmatrix} -1 & & \\ & -1 & \\ & & 1 \end{pmatrix}$, therefore this matrix is mapped onto the unit matrix $1^{(8)} \otimes 1^{(3)}$ by transition to the laboratory system. The group $\overline{\mathcal{H}}(\tau)$ is identical with the "double group" of the Longuet-Higgins group of this SRM given in a recent paper by Merer et al.[23)].

3 Applications of Isometric Groups

In this chapter applications of isometric groups will be presented in the sequence:
3.1. Isometric group and Born-Oppenheimer Approximation.
3.2. Rotation-large amplitude internal motion problem of SRMs and isometric groups.
3.3. Irreducible tensors and selection rules of SRMs.
3.4. Chirality of nonrigid molecules.
3.5. Enumeration and classification of conformational isomers of nonrigid molecules.

3.1 Isometric Group and Born-Oppenheimer Approximation

In this section some aspects of the symmetry of the molecular Schrödinger operator[7]

$$\hat{H} = \hat{T}_n + \hat{H}^0 \tag{3.1}$$

$$\hat{H}^0(p_j, x_j; X_k(\xi)) = \sum_{j=1}^{N} \left\{ \frac{p_j^2}{2\,m_0} - \sum_{k=1}^{K} \frac{e_0^2 Z_k}{|x_j - X_k(\xi)|} \right\} + \sum_{j,j'}^{j<j'} \frac{e_0^2}{|x_j - x_{j'}|} + \sum_{k,k'}^{k<k'} \frac{e_0^2 Z_k Z_{k'}}{|X_k(\xi) - X_{k'}(\xi)|} \tag{3.2}$$

$$\hat{T}_n = \sum_{k=1}^{K} \frac{P_k^2}{2\,M_k} \tag{3.3}$$

will be considered. The original Born-Oppenheimer approximation consists in the solution of the eigenvalue problem

7 In this section the symbol $\hat{H}$ is used for the molecular Schrödinger operator.

$$\hat{H}^0(p_j, x_j; X_k(\xi))\, \psi(x_j, s_{j3}; X_k(\xi)) = \epsilon^0(X_k(\xi))\, \psi(x_j, s_{j3}; X_k(\xi)) \qquad (3.2')$$

for a continuous set of nuclear configurations, whereby the nuclear coordinates are considered as parameters (not operators). The symmetry group of the operator (3.1) is the direct product of the inhomogeneous 3-dimensional orthogonal group IO(3) (whose operators are to be applied cogrediently to electron and nuclear dynamical variables), the symmetric group γ_N of the permutation of the indices of the electrons, the direct product of the symmetric groups γ_{K_s} of the indices of the nuclei of the sets of identical nuclei, i.e. $\otimes_s \gamma_{K_s}$, and the time reversal group[29]. For applications to molecular physics and chemistry the symmetry of the operator (3.2) is more relevant, at least as far as chemical reactions are excluded, therefore, this symmetry will exclusively be discussed here. The following statements may immediately be made concerning the symmetry group of $\hat{H}^0$: $\hat{H}^0$ is symmetric w.r.t.

(i) the symmetric group of the permutation of the indices of the electrons

(ii) the covering group $\mathscr{G}(\xi)$ of the NC $\{X_k(\xi), Z_k, M_k\}$, more precisely, the covering group of the NC $\{X_k(\xi), Z_k\}$, since the nuclear masses do not show up in the operator $\hat{H}^0$. The group $\mathscr{G}(\xi)$ is isomorphic to a permutation group $\Pi\{\mathscr{G}\}$ of the nuclei of a set of equivalent nuclei, which is a subgroup of the symmetric group of the set.

The group $\mathscr{G}(\xi)$ implies, that the operator $\hat{H}^0$ is symmetric w.r.t. to the transformations of the dynamical variables of the electrons under a group $\mathscr{G}^{(el)}$

$$\mathscr{G}^{(el)} \stackrel{is}{=} \mathscr{G}(\xi) \qquad (3.4)$$

(iii) the group IO(3), applied cogrediently to the electron dynamical variables and the nuclear coordinate vectors. However, application of the operators of this group to either the nuclear coordinates alone or to the electron dynamical variables alone does not in general leave the operator $\hat{H}^0$ symmetric.

From the statements (ii) and (iii) one obtains

$$P_G\epsilon^0(X_k(\xi)) = \epsilon^0(X_k(\xi)) \qquad (3.5)$$

As a next step we consider the transformation properties of $\hat{H}^0$ w.r.t. $\mathscr{F}(\xi)$, provided the operators $\hat{P}_F$ are applied to the nuclear and electron coordinates according to the following propositions, $\forall F \in \mathscr{F}(\xi)$:

$$\begin{aligned} &\hat{P}_F\{\widetilde{X}_k(\xi)\} = \{\widetilde{X}_k(\xi)\}\cdot\Gamma^{(NCf)}(F) = \{\widetilde{X}_k(\xi)\}\Pi(F)\otimes\Gamma^{(3)}(F)\\ &\hat{P}_F(\hat{x}_j)\hat{P}_F^{-1} = \widetilde{\Gamma}^{(3)}(F)(\hat{x}_j)\\ &\hat{P}_F(\hat{p}_j)\hat{P}_F^{-1} = \widetilde{\Gamma}^{(3)}(F)(\hat{p}_j)\end{aligned} \qquad (3.6)$$

Obviously according to (ii) and (iii)

$$\hat{P}_F\hat{H}^0\hat{P}_F^{-1} = \hat{H}^0 \qquad (3.7)$$

It should be pointed out that the operators $\hat{P}_F$ which by definition act only on internal coordinates, are not symmetries of $\hat{H}^0$:

$$\hat{P}_F\hat{H}^0\hat{P}_F^{-1} = \hat{H}^0(p_j, x_j; \widetilde{\Gamma}^{(3)}(F)X_{Fk}(\xi))$$
$$= \sum_j\left\{\frac{p_j^2}{2\,m_0} - \sum_k e_0^2 Z_k\,|x_j - \widetilde{\Gamma}^{(3)}(F)X_{Fk}(\xi)|^{-1}\right\} + \sum_{j<j'}\frac{e_0^2}{|x_j - x_{j'}|}$$
$$+ \sum\sum_{kk'} e_0^2 Z_k Z_{k'}\,|\widetilde{\Gamma}^{(3)}(F)X_{Fk}(\xi) - \widetilde{\Gamma}^{(3)}(F)X_{Fk'}(\xi)|^{-1} \neq \hat{H}^0 \tag{3.8}$$

Next we state the important theorem: $\forall\, F \in \mathscr{F}(\xi)$

$$\hat{P}_F\epsilon^0(X_k(\xi)) = \epsilon^0(X_k(F^{-1}(\xi))) = \epsilon^0(X_k(\xi)) \tag{3.9}$$

Again this relation follows from the symmetries (ii) and (iii); it expresses that the electronic energy function assumes the same value for all isometric NCs. Equations (3.5) and (3.9) show that $\epsilon^0(X_k(\xi))$ is symmetric w.r.t. to the full isometric group. Whereas the symmetry of $\epsilon^0(X_k(\xi))$ w.r.t. $\mathscr{G}(\xi)$ merely expresses that $\epsilon^0(X_k(\xi))$ is a function of the internal coordinates only, its symmetry w.r.t. $\mathscr{F}(\xi)$ is a genuine symmetry.

3.2 Rotation-Large Amplitude Internal Motion Problem of SRMs

The classical energy function of a SRM may be written in the form

$$T + V = \frac{1}{2}(\widetilde{\omega\dot{\xi}})(g_{mn}(\xi))\begin{pmatrix}\omega\\ \dot{\xi}\end{pmatrix} + V(\xi) \tag{3.10}$$

In this equation the angular velocity vector ω is referred to the center of mass frame coordinate system and the $\dot{\xi}$s are the time derivatives of the internal coordinates. Moreover, the kinetic energy matrix coefficients may be expressed as

$$g_{pq} = I_{pq} = \sum_k M_k[\widetilde{X}_k(\xi)X_k(\xi)\delta_{pq} - X_{kp}(\xi)X_{kq}(\xi)]$$

$p, q = 1, 2, 3$

$$g_{pr} = \sum_k M_k\left[X_k(\xi), \frac{\partial X_k(\xi)}{\partial \xi_r}\right]_p, \quad r = 1, 2, \ldots, f$$

$$g_{rs} = \sum_k M_k\left(\frac{\partial X_k(\xi)}{\partial \xi_r}, \frac{\partial X_k(\xi)}{\partial \xi_s}\right), \quad r, s = 1, 2, \ldots, f \tag{3.11}$$

The potential $V(\xi)$ may now be identified with the electronic energy function

$$V(\xi) = \epsilon^0(X_k(\xi)) \tag{3.12}$$

provided the interaction of nuclear and electronic motion may be neglected, i.e. the adiabatic approximation is adequate.

If in place of the nonholonomic coordinates ω the eulerian angular velocities are introduced by[30)]

$$\omega = \begin{bmatrix} -\sin\beta\cos\gamma & \sin\gamma & 0 \\ \sin\beta\sin\gamma & \cos\gamma & 0 \\ \cos\beta & 0 & 1 \end{bmatrix} \begin{bmatrix} \dot\alpha \\ \dot\beta \\ \dot\gamma \end{bmatrix} = E(\epsilon)\dot\epsilon \;^{8} \tag{3.13}$$

Equation (3.10) goes over into the lagrangian form

$$T = \frac{1}{2}(\widetilde{\dot\epsilon\dot\xi}) \begin{bmatrix} \tilde E(\epsilon) & 0 \\ 0 & 1^{(f)} \end{bmatrix} (g_{mn}(\xi)) \begin{bmatrix} E(\epsilon) & 0 \\ 0 & 1^{(f)} \end{bmatrix} \begin{bmatrix} \dot\epsilon \\ \dot\xi \end{bmatrix} \tag{3.10'}$$

and the corresponding hamiltonian form

$$T = \frac{1}{2}(\tilde p_\epsilon \tilde p_\xi) \begin{bmatrix} E^{-1}(\epsilon) & 0 \\ 0 & 1^{(f)} \end{bmatrix} (g^{mn}(\xi)) \begin{bmatrix} \tilde E^{-1}(\epsilon) & 0 \\ 0 & 1^{(f)} \end{bmatrix} \begin{bmatrix} p_\epsilon \\ p_\xi \end{bmatrix} \tag{3.10''}$$

The total angular momentum referred to the frame system is given by

$$\tilde P = \tilde p_\epsilon E^{-1}(\epsilon) \tag{3.14}$$

and if this is used, the kinetic energy becomes

$$T = \frac{1}{2}(\widetilde{P p_\xi})(g^{mn}(\xi)) \begin{bmatrix} P \\ p_\xi \end{bmatrix} \tag{3.10'''}$$

3.2.1 Symmetry Group of the Hamiltonian

The foregoing discussion allows to state the theorem: The full isometric group $\mathscr{H}(\xi)$ is a proper or improper subgroup of the symmetry group $\overline{\mathscr{G}}\{\hat H\}$ of the rotation internal motion hamiltonian $\hat H = \hat T + \hat V$

$$\hat P_H \hat H \hat P_H^{-1} = \hat H, \quad \forall H \in \mathscr{H}(\xi) \tag{3.15}$$

Since the dynamical problem (3.10) refers to the LS, the primitive period isometric transformations are to be included in $\mathscr{H}(\xi)$. A proof of this important theorem has been given earlier[14)] [9]. $\overline{\mathscr{G}}\{\hat H\}$ represents symmetry of $\hat H$ w.r.t. to operations of the

8 For the sake of brevity the set $\alpha\beta\gamma$ will often be abbreviated by ϵ and the velocity vector $\begin{bmatrix} \dot\alpha \\ \dot\beta \\ \dot\gamma \end{bmatrix}$ by $\dot\epsilon$.

9 In the Eqs. (5.19) and (5.22) of this paper the matrix $\Gamma^{(3)}(F)$ should be replaced by $R(F) = |\Gamma^{(3)}(F)| \cdot \Gamma^{(3)}(F)$.

frame system. On the other hand $\hat{H}$ is symmetric w.r.t. arbitrary orthogonal transformations of the LS

$$\hat{P}_{R^l}\hat{H}\hat{P}_{R^l}^{-1} = \hat{H} \tag{3.16}$$

The set of properly orthogonal transformations R^l forms the group $SO(3)^l$, the reflexion Z^l at the origin of the LS likewise leaves $\hat{H}$ symmetric, since the eulerian angles remain unaffected by Z^l. Therefore, $\hat{H}$ is symmetric w.r.t. the full rotation group $O(3)^l$. However, in agreement with the usual conventions we will omit the elements $Z^lR^l \in O(3)^l$. As a consequence we will consider hence-forward the group

$$SO(3)^l \otimes \overline{\mathscr{G}}\{\hat{H}\} \tag{3.17}$$

as the symmetry group $\mathscr{G}\{\hat{H}\}$ of the rotation internal motion problem.

For a number of SRMs $\mathscr{H}(\xi) \subset \overline{\mathscr{G}}\{H\}$, i.e. $\mathscr{H}(\xi)$ is a proper subgroup of $\overline{\mathscr{G}}\{\hat{H}\}$. This fact has been established by determining the group $\overline{\mathscr{G}}\{\hat{H}\}$ directly. In the case where a principle axis coincides with an internal rotation axis for all values of the internal rotation angle, but not being a covering symmetry axis, we have

$$\mathscr{H}(\xi) \subset \overline{\mathscr{G}}\{\hat{H}\} \tag{3.18}$$

$\overline{\mathscr{G}}\{\hat{H}\}$ now contains an element with the structure

$$\Gamma(R) = \begin{bmatrix} B(R) & 0 & b(R) \\ \cdot & 1^{(f)} & 0 \\ \cdot & \cdot & 1 \end{bmatrix}$$

not contained in $\Gamma\{\mathscr{H}\}$. A typical case is the SRM $C_{2v}F\ C_{3v}T$, which might serve as a model for the rotation-internal rotation problem of nitromethane type molecules. A further case is given by SRMs whose group $\mathscr{G}(\xi)$ contains an element of order $\geqslant 3$. For this case $\overline{\mathscr{G}}\{\hat{H}\}$ is an infinite group since it contains now an infinitesimal rotation represented by ($\delta \in [0, 2\pi]$ being an arbitrary angle)

$$B(R) = \begin{bmatrix} 1 & & \\ & 1 & \\ & & 1 \end{bmatrix}, \quad b(R) = \begin{bmatrix} 0 \\ 0 \\ \delta \end{bmatrix} \tag{3.19}$$

3.2.2 Solution of the Energy Eigenvalue Problem

The energy eigenvalue problem associated with the classical energy function (3.10) is

$$\hat{H}\,\psi = E\,\psi \tag{3.20}$$

where the energy operator is obtained from Eq. (3.10″) by the usual procedure[31, 32] leading to

$$\hat{H} = \frac{1}{2}(\tilde{\hat{P}}, g^{-\frac{1}{4}}\tilde{\hat{p}}_\xi g^{\frac{1}{4}})(g^{mn}(\xi)) \begin{bmatrix} \hat{P} \\ g^{\frac{1}{4}}\hat{p}_\xi g^{-\frac{1}{4}} \end{bmatrix} + \hat{V}(\xi) \tag{3.21}$$

The eigenvalue problem Eq. (3.20) has to be solved in the Hilbert space $\mathbb{H}\{H\}$ defined by

$$\begin{aligned} \mathbb{H}\{\hat{H}\} := \{f(\epsilon, \xi) \mid (f, f) &= \int f^* f \sin\beta \, d\alpha d\beta d\gamma d^f\xi < \infty \\ (f, g) &= \int f^* g \sin\beta \, d\alpha d\beta d\gamma d^f\xi\} \end{aligned} \tag{3.22}$$

and the function space (domain of $\hat{H}$)

$$\begin{aligned} \mathbb{A}\{\hat{H}\} := \{&u(\epsilon, \xi) \mid u(\epsilon, \xi) \in \mathbb{C}^{(2)}[0 \leqslant \alpha < 2\pi, 0 \leqslant \beta < \pi, 0 \leqslant \gamma < 2\pi, \xi \in \mathbb{D}(\xi)] \\ &(u, v) = \int u^* v \sin\beta \, d\alpha \, d\beta \, d\gamma \, d^f\xi \\ &\hat{H}u \in \mathbb{H}\{\hat{H}\}\} \end{aligned} \tag{3.23}$$

The energy eigenfunctions may be classified according to the irreducible representations of the group $\underline{SO(3)^I \otimes \overline{\mathscr{G}}}\{\hat{H}\}$. For the sake of simplicity we will assume $\overline{\mathscr{G}}\{\hat{H}\} \overset{is}{=} \mathscr{H}(\xi)$; in cases, where Eq. (3.18) is valid, an appropriate extension of $\mathscr{H}(\xi)$ has to be used. Denoting the irreducible representations of the group (3.17) by

$$D^{(J)} \otimes \Gamma^{(j)} \tag{3.24}$$

the eigenvalue problem (3.20) may be specified as

$$\hat{H}\psi_{JM\Gamma^{(j)}\mu N}(\epsilon, \xi) = E_{J\Gamma^{(j)}N}\,\psi_{JM\Gamma^{(j)}\mu N}(\epsilon, \xi) \tag{3.20'}$$

where M and μ denote row indices of $D^{(J)}$ and $\Gamma^{(j)}$, respectively[33]. For solution of Eq. (3.20) one usually uses a matrix representation of $\hat{H}$ in a suitable chosen zeroth order basis. A practical choice is the direct product of the set of rotation group coefficients[34]

$$\{D^{(J)}_{M'M}(\epsilon)\} := \{D^{(J)}_{M'M}(\alpha\beta\gamma) \mid J \in \mathbb{N};\; M', M \in [-J, +J]\} \tag{3.25}$$

and suitably chosen complete orthonormal bases

$$\begin{aligned} \{\varphi^{(k)}_m(\xi_k)\} := \{&\varphi^{(k)}_m(\xi) \mid \varphi^{(k)}_m(\xi_k) \in \mathbb{C}^{(2)}[\xi_k \in \mathbb{D}_k], m \in \mathbb{N} \\ &(\varphi^{(k)}_{\overline{m}}, \varphi^{(k)}_m) = \int \varphi^{(k)*}_{\overline{m}} \varphi^{(k)}_m d\xi_k \\ &\| \varphi^{(k)}_m \|^2 = \int |\varphi^{(k)}_m|^2 d\xi_k < \infty\} \end{aligned}$$

e.g. Fourier functions for angular internal coordinates or Laguerre or Hermite functions for distances.

In practical applications one may first solve the "internal problem" defined by the operator

$$\hat{H}^{int} = \frac{1}{2}\left(g^{-\frac{1}{4}}\tilde{\hat{p}}_{\xi} g^{\frac{1}{4}}\right)(g_{int}^{mn})\left(g^{\frac{1}{4}}\hat{p}_{\xi} g^{-\frac{1}{4}}\right) + \hat{V}(\xi) \qquad (3.21')$$

which defines all states with J = 0, i.e. vanishing overall angular momentum[35]. Typical examples have been given in the literature[36–38, 25]. It should be pointed out that the eigenstates of the internal problem may, according to the definition of the isometric group, already uniquely be classified according to the irreducible representations of the group $\mathscr{F}(\xi)$. The extension of $\mathscr{F}(\xi)$ to $\mathscr{H}(\xi)$ by the covering group $\mathscr{G}(\xi)$ leads to a refinement of the group theoretical classification of the internal eigenstates $u_{\Gamma^{(j)}\mu N}(\xi)$. Since all such states have to be symmetric w.r.t. the operators P_G, the $u_{\Gamma^{(j)}\mu N}(\xi)$ belong exclusively to those irreducible representations of $\mathscr{H}(\xi)$, which belong to the normal subgroup $\mathscr{G}(\xi)$ of $\mathscr{H}(\xi)$. Typical examples where this situation is realized are SRMs of the types $C_{2v}F{-}C_{2v}I$[37] (N≡C–NH_2, $\overline{CH_2CH_2NH}$), D_{2h}/D_{4h} ring puckering[38] (cyclobutane) and $D_{\infty h}F(C_ST)_2$[25] (glyoxal, H_2O_2).

3.2.2.1 Symmetrization of the Zeroth Order Basis. By subjecting the elements of the direct product basis

$$\{D^{(J)}_{M'M}(\epsilon)\prod_k \varphi^{(k)}_m(\xi_k)\}$$

to the transformations of $\Gamma\{\mathscr{H}\}$, i.e. to the operators $\hat{P}_H$ associated with the transformations Eq. (2.83), one may construct by well known procedure[39] zeroth order basis functions belonging to the irreducible representation $D^{(J)} \otimes \Gamma^{(j)}$. The use of symmetrized basis functions leads to considerable simplifications in practical solutions of the energy eigenvalue problem.

In the symmetrization process the primitive period isometric transformations F_p play an outstanding role. For all SRMs with $\Gamma^{(3)}(F_p) \in SO(3)$ these operators are represented in the representation $\Gamma^{(NCI)}\{\mathscr{H}\}$ by the unit matrix (cf. Sect. 2.3.3). This implies for such SRMs the important relation

$$\hat{P}_{F_p}\psi_{JM\Gamma^{(j)}\mu N}(\epsilon,\xi) = \psi_{JM\Gamma^{(j)}\mu N}(\epsilon,\xi),\ \forall F_p \in \overline{\mathscr{F}}(\xi)$$
$$\forall\, \psi_{JM\Gamma^{(j)}\mu N} \in \sigma\{\hat{H}\} \qquad (3.26)$$

i.e. the energy eigenfunctions are symmetric w.r.t. the operators $\hat{P}_{F_p}$. The latter therefore on one hand imply periodicity conditions for the energy eigenfunctions and symmetrized (zeroth order) basis functions and on the other hand cause the energy eigenfunctions to belong uniquely to the irreducible representations $\Gamma^{(j)}\{\mathscr{H}\}$ of the group $\mathscr{H}(\xi)$.

For all SRMs with $\overline{\mathscr{H}}(\xi) \overset{end}{=} \mathscr{H}(\xi)$ the interrelation of the representations $\Gamma^{(j)}\{\mathscr{H}\}$ and $\Gamma^{(j)}\{\overline{\mathscr{H}}\}$ may be investigated by means of the Frobenius theorem on induced representations[40]. If in addition $\mathscr{H}(\xi)$ is a normal subgroup of $\overline{\mathscr{H}}(\xi)$, as it has actually been the case in most examples studied up to now, the irreducible representations $\Gamma^{(j)}\{\overline{\mathscr{H}}\}$, which occur as symmetries of the energy eigenfunctions follow uniquely from $\Gamma^{(j)}\{\mathscr{H}\}$ by means of the representations of the factor group $\overline{\mathscr{H}}(\xi)/\mathscr{H}(\xi)$. The group theoretical relations arising from the primitive period isometric substitutions form the mathematical background for the introduction of "double groups" by several authors[23, 24, 27] as has already been pointed out in Sect. 2.3.3. We shall not distinguish between $\mathscr{H}$ and $\overline{\mathscr{H}}$ till further notice, but as shown in the discussion above primitive period isometric transformations have to be included when considering the symmetry group of the operator (3.21).

3.3 Irreducible Tensors and Selection Rules for SRMs

In this section we first present a set of general transformation formulae for tensor operators associated with SRMs. These then serve as a mathematical tool for the formulation of Wigner-Eckart theorems and selection rules for irreducible tensor operators associated with multipole transitions of SRMs. The concept of isometric groups will allow a formulation of selection rules in strict analogy to the group theoretical treatment of quasirigid molecules first presented by Wigner[5].

3.3.1 Transformation Properties of Tensors w.r.t. Isometric Transformations

The following treatment is based on the assumptions

(i) transformation formula for a polar vector (operator) expressed in cartesian coordinates w.r.t. $\hat{P}_H \in \mathscr{H}(\xi)$

$$\hat{P}_H(V_k^f(\xi))\hat{P}_H^{-1} = (V_k^f(H^{-1}(\xi))) = \widetilde{\Gamma}^{(3)}(H)(V_k^f(\xi)), \quad \forall H \in \mathscr{H}(\xi) \tag{3.27}$$

Typical examples for vectors of this type are the electrical dipole moment $\mathbf{M}^{(e)f}(\xi)$ or any vector of the type $\mathbf{X}^f(\xi) = \sum_k Z_k \mathbf{X}_k^f(\xi)$, where the sum is to be extended over a set of nuclei equivalent w.r.t. $\mathscr{H}(\xi)$. Equation (3.27) follows immediately from the properties of $\Gamma^{(NCf)}\{\mathscr{H}\}$ and has been proved earlier for the expectation value of the electric dipole operator

$$\hat{\mu}^{(e)f} = -e_0 \sum_j \hat{x}_j + e_0 \sum_k Z_k \mathbf{X}_k(\xi)$$

for molecules in a given electronic state[41].

(ii) Transformation formula for an axial vector (vector operator):

$$\hat{P}_H(V_k^f(\xi))\hat{P}_H^{-1} = (V_k^f(H^{-1}(\xi))) = \widetilde{R}(H)(V_k^f(\xi)), \quad \forall H \in \mathscr{H}(\xi) \tag{3.28}$$

typical examples being the magnetic dipole operator $\hat{\mathbf{M}}^{(m)f}(\xi)$, the angular momentum w.r.t. $\widetilde{\mathbf{e}}^f$, etc.

(iii) The coupling operator for a tensor T^f of rank n for a SRM in a multipole electric or magnetic field $E^{1\otimes n}$ ($\otimes$n denotes the nth Kronecker power)

$$\hat{H}' = -\widetilde{E}^{l\otimes n}\widetilde{R}^{\otimes n}(\epsilon)T^f(\xi) \qquad (3.29)$$

where the tensor $T^f(\xi)$ transforms according to

$$\hat{P}_H T^f(\xi)\hat{P}_H^{-1} = T^f(H^{-1}(\xi)) = \widetilde{\Gamma}^{(3)\otimes n}(H)\cdot T^f(\xi) \qquad (3.30)$$

$$\hat{P}_H T^f(\xi)\hat{P}_H^{-1} = \widetilde{R}^{\otimes n}(H)\cdot T^f(\xi) \qquad (3.30')$$

$$\hat{P}_H T^f(\xi)\hat{P}_H^{-1} = \widetilde{R}^{\otimes n_1}(H)\otimes\widetilde{\Gamma}^{(3)\otimes n_2}(H)T^f(\xi),\ \forall H\in\mathscr{H}(\xi)$$

$$n_1 + n_2 = n \qquad (3.30'')$$

whether or not $T^f(\xi)$ transforms w.r.t. $\hat{P}_H$ as $M^{(e)f\otimes n}$, $M^{(m)f\otimes n}$ or $M^{(m)f\otimes n_1}\otimes M^{(e)f\otimes n_2}$.

The coupling operator of an irreducible operator $(\hat{A}^f_{s\sigma}(\xi))$ of rank s for a semirigid model in a multipole field $(E^l_{s\sigma})$ may be taken in the form

$$\hat{H}' = -(E^l_{s\sigma})^\dagger D^{(s+)}(\epsilon)^\dagger(\hat{A}^f_{s\sigma}(\xi)) \qquad (3.31)$$

where $D^{(s+)}(\epsilon)^\dagger(\hat{A}^f_{s\sigma}(\xi)) = (\hat{A}^l_{s\sigma}(\epsilon,\xi))$, (3.32)

expresses the tensor $\hat{A}$ in the laboratory system.

Further important mathematical tools are the two fundamental formulae from rotation group theory ($s\in\mathbb{N}$)

$$\hat{P}_{R^l}D^{(s)}(\epsilon) = D^{(s)}(\epsilon)D^{(s)}(R^l),\ \forall R^l\in SO(3)^l \qquad (3.33)$$

$$\hat{P}_{R^f}D^{(s)}(\epsilon) = D^{(s)}(R^f)^\dagger D^{(s)}(\epsilon),\ \forall R^f\in SO(3)^f \qquad (3.33')^{10}$$

Eqs. (3.33) and (3.33′) express the transformation behavior of the rotation group matrices w.r.t. rotations of the laboratory system $\widetilde{e}^l$ and the frame system $\widetilde{e}^f$, respectively. The second Eq. (3.33′) assumes for the case $H\in\mathscr{H}(\xi)$ the special form

$$\hat{P}_H D^{(s+)}(\epsilon) = D^{(s+)}(H)^\dagger\cdot D^{(s+)}(\epsilon) \qquad (3.33'')$$

A proof of this equation is given in Appendix 2.

Basing on Eqs. (3.27), (3.28) and (3.30)

$$\hat{P}_H(\hat{A}^f_{s\sigma}(\xi))\hat{P}_H^{-1} = (\hat{A}^f_{s\sigma}(H^{-1}(\xi))) = D^{(sp)}(H)^\dagger(\hat{A}^f_{s\sigma}(\xi)) \qquad (3.34)$$

where p denotes the parity of the tensor operator $\hat{A}^f$.

Furthermore, from Eqs. (3.33) and (3.34) we find the following important transformation formula for the coupling operator ($\forall R^l\in SO(3)^l$, $\forall H\in\mathscr{H}(\xi)$)

$$\hat{P}_{R^l}\hat{P}_H\hat{H}'\hat{P}_H^{-1}\hat{P}_{R^l}^{-1}$$

$$= -(E^l_{s\sigma})^\dagger D^{(s+)}(R^l)^\dagger D^{(s+)}(\epsilon)^\dagger D^{(s+)}(H)D^{(sp)}(H)^\dagger(\hat{A}^f_{s\sigma}(\xi))$$

$$= \begin{cases} -(E^l_{s\sigma})^\dagger D^{(s+)}(R^l)^\dagger D^{(s+)}(\epsilon)^\dagger(\hat{A}^f_{s\sigma}(\xi)), \text{ if } p=+1 \text{ (even)} \\ -(E^l_{s\sigma})^\dagger D^{(s+)}(R^l)^\dagger D^{(s+)}(\epsilon)^\dagger\,|\Gamma^{(3)}(H)|\,(\hat{A}^f_{s\sigma}(\xi)), \text{ if } p=-1 \text{ (odd)} \end{cases} \qquad (3.35)$$

10 The formula (3.13) given in Ref.[15] should be corrected according to Eq. (3.33′).

For electric and magnetic dipole moment and the electric and magnetic quadrupole operator the last equation yields

$$\hat{P}_{R^l}\hat{P}_H\hat{H}'\hat{P}_H^{-1}\hat{P}_{R^l}^{-1} = -(E^l_\sigma)^\dagger D^{(1+)}(R^l)^\dagger D^{(1+)}(\epsilon)^\dagger\,|\,\Gamma^{(3)}(H)\,|\,(\hat{M}^{(e)f}_\sigma(\xi))$$

$$= -(E^l_\sigma)^\dagger D^{(1+)}(R^l)^\dagger D^{(1+)}(\epsilon)^\dagger T\,|\,\Gamma^{(3)}(H)\,|\,(\hat{M}^{(e)f}_k(\xi)) \quad (3.36)$$

$$\hat{P}_{R^l}\hat{P}_H\hat{H}'\hat{P}_H^{-1}\hat{P}_{R^l}^{-1} = -(E^l_\sigma)^\dagger D^{(1+)}(R^l)^\dagger D^{(1+)}(\epsilon)^\dagger T(\hat{M}^{(m)f}_k(\xi)) \quad (3.36')$$

$$\hat{P}_{R^l}\hat{P}_H\hat{H}'\hat{P}_H^{-1}\hat{P}_{R^l}^{-1} = -(E^l_{2\sigma})^\dagger D^{(2+)}(R^l)^\dagger D^{(2+)}(\epsilon)^\dagger(\hat{A}^f_{2\sigma}(\xi)) \quad (3.36'')$$

$$\text{where } T = \begin{bmatrix} \frac{1}{\sqrt{2}} & \frac{i}{\sqrt{2}} & 0 \\ 0 & 0 & 1 \\ -\frac{1}{\sqrt{2}} & \frac{i}{\sqrt{2}} & 0 \end{bmatrix}$$

The relation between the spherical components $\hat{A}^f_{2\sigma}(\xi)$ of a general tensor $\hat{A}^f_2$ of rank 2 and the cartesian components $\hat{A}^f_{mn}(\xi)$ are given in Appendix 4. Equations (3.36) will form the basis for derivation of selection rules for rotation-internal motion transitions of SRMs presented in the next section. They also may serve for derivation of the transformation properties of the electric and magnetic dipole moment operators referred to the laboratory system ($\forall H \in \mathscr{H}(\xi)$):

$$\hat{P}_H(\hat{M}^{(e)l}_k(\epsilon,\xi))\hat{P}_H^{-1} = |\,\Gamma^{(3)}(H)\,|\,(\hat{M}^{(e)l}_k(\epsilon,\xi)) \quad (3.37)$$

$$\hat{P}_H(\hat{M}^{(m)l}_k(\epsilon,\xi))\hat{P}_H^{-1} = (\hat{M}^{(m)l}_k(\epsilon,\xi)) \quad (3.37')$$

Analogously for a tensor operator of rank 2 with even parity

$$\hat{P}_H(\hat{A}^l_{2\sigma}(\epsilon,\xi))\hat{P}_H^{-1} = (\hat{A}^l_{2\sigma}(\epsilon,\xi)) \quad (3.37'')$$

The simplicity of these transformation formulae is to be traced back to the general formulae (3.33), (3.34) and the fact that the operators $\hat{P}_H$ act on both the eulerian angles and the internal coordinates simultaneously, as expressed by the representation $\Gamma\{\mathscr{H}\}$. The analogy of the Eqs. (3.37) to the representation $\Gamma^{(NCI)}\{\mathscr{H}\}$ should be noted.

3.3.2 Wigner-Eckart Theorem and Selection Rules

1 Electric Dipole Transitions

Derivation of the Wigner-Eckart theorem (WET) will be based on the transformation properties of the energy eigenfunctions expressed by

$$\hat{P}_{R^l}\hat{P}_H\psi_{JM\Gamma^{(j)}\mu N}(\epsilon,\xi) = \sum_{M'}\sum_{\mu'}\psi_{JM'\Gamma^{(j)}\mu' N} D^{(J)}_{M'M}(R^l)\cdot\Gamma^{(j)}_{\mu'\mu}(H) \quad (3.38)$$

Formulation of the matrix elements of the coupling operator (3.31) yields

$$(\psi_{\bar{J}\bar{M}\Gamma^{(\bar{j})}\bar{\mu}\bar{N}}(\epsilon, \xi), \hat{H}'\psi_{JM\Gamma^{(j)}\mu N}(\epsilon, \xi))$$
$$= (\hat{P}_{R^1}\hat{P}_H\psi_{\bar{J}\bar{M}\Gamma^{(\bar{J})}\bar{\mu}\bar{N}}, \hat{P}_{R^1}\hat{P}_H\hat{H}'\hat{P}_H^{-1}\hat{P}_{R^1}^{-1}\hat{P}_{R^1}\hat{P}_H\psi_{JM\Gamma^{(j)}\mu N}(\epsilon, \xi)) \tag{3.39}$$

Since the matrix elements of $\hat{H}'$ may be obtained from those of $(\hat{M}^{(e)1}(\epsilon, \xi))$ by linear combination, the latter will be calculated first:

$$(\psi_{\bar{J}\bar{M}\Gamma^{(\bar{J})}\bar{\mu}\bar{N}}, (\hat{M}_\sigma^{(e)1})\psi_{JM\Gamma^{(j)}\mu N})$$
$$= \sum_{\bar{M}'}\sum_{M'}\sum_{\bar{\mu}'}\sum_{\mu'} D^{(\bar{J})}_{\bar{M}'\bar{M}}(R^1)^* D^{(J)}_{M'M}(R^1)\Gamma^{(\bar{j})}_{\bar{\mu}'\bar{\mu}}(H)^*\Gamma^{(j)}_{\mu'\mu}(H)|\Gamma^{(3)}(H)|$$
$$\times\ D^{(1)}(R^1)^\dagger(\psi_{\bar{J}\bar{M}'\Gamma^{(\bar{j})}\bar{\mu}'\bar{N}}(\epsilon, \xi), (\hat{M}_\sigma^{(e)1}(\epsilon, \xi))\psi_{JM'\Gamma^{(j)}\mu'N}(\epsilon, \xi))$$

Integration over $SO(3)^1$ and summation over $\mathscr{H}(\xi)$ yields the vector of matrix elements ($h = |\mathscr{H}(\xi)|$)

$$(\psi_{\bar{J}\bar{M}\Gamma^{(\bar{J})}\bar{\mu}\bar{N}}, (\hat{M}_\sigma^{(e)1})\psi_{JM\Gamma^{(j)}\mu N})$$
$$= \sum_{\bar{M}'}\sum_{M'}(2J+1)^{-1}(S_{JM'\bar{J}\bar{M}'1\bar{\sigma}}S_{JM\bar{J}\bar{M}1\sigma})$$
$$\times \sum_{\bar{\mu}'}\sum_{\mu'}(\psi_{\bar{J}\bar{M}'\Gamma^{(\bar{J})}\bar{\mu}'\bar{N}}, (\hat{M}_\sigma^{(e)1})\psi_{JM'\Gamma^{(j)}\mu'N})$$
$$\times\ h^{-1}\sum_{\mathscr{H}}\Gamma^{(\bar{J})}_{\bar{\mu}'\bar{\mu}}(H)^*\Gamma^{(j)}_{\mu'\mu}(H)|\Gamma^{(3)}(H)| \tag{3.40}$$

The righthand side contains the Wigner coefficient $S_{JM\bar{J}\bar{M}1\sigma}$, $\sigma = -1, 0, +1$, which expresses the usual rotation group selection rules

$$\Delta J = 0, \pm 1, \qquad J = 0 \not\leftrightarrow J = 0$$
$$\Delta M = 0 \text{ for z-polarization } (\sigma = 0) \tag{3.41}$$
$$\Delta J = 0, \pm 1, \qquad J = 0 \not\leftrightarrow J = 0$$
$$\Delta M = \pm 1 \text{ for x, y-polarization } (\sigma = \pm 1) \tag{3.41'}$$

and the quantity

$$\Theta^{(\bar{J}j)}_{\bar{\mu}\mu} = h^{-1}\sum_{\mathscr{H}}\Gamma^{(\bar{J})}_{\bar{\mu}'\bar{\mu}}(H)^*\Gamma^{(j)}_{\mu'\mu}(H)|\Gamma^{(3)}(H)| \tag{3.42}$$

which expresses the selection rules w.r.t. the isometric group $\mathscr{H}(\xi)$.

The two Eqs. (3.40) and (3.42) may be discussed as follows:

(i) the selection rules w.r.t. $\mathscr{H}(\xi)$ are the same for all components (cartesian or spherical).

(ii) the quantity $\Theta^{(\bar{J}j)}_{\bar{\mu}\mu}$ may be specified further for the case a, b1, b2 defined in Sect. 2.2:

Case a:

$$|\Gamma^{(3)}(H)| = 1, \ \forall H \in \mathscr{H}(\xi)$$

$$\Theta^{(\bar{j}j)}_{\bar{\mu}\mu} = l_j^{-1}\delta_{\bar{j}j}\,\delta_{\bar{\mu}'\mu'}\delta_{\bar{\mu}\mu} \qquad (3.42a)$$

where the symbol $\delta_{\bar{\mu}'\mu'}$ does not express a selection rule, since $\bar{\mu}'$, μ' are subject to the summation in Eq. (3.40). l_j denotes the dimension of $\Gamma^{(j)}$. Hence the electric dipole selection rule reads

$$\Gamma^{(j)} \longleftrightarrow \Gamma^{(j)}$$

$$\mu \quad \longleftrightarrow \mu$$

Case b:

The group $\Gamma^{(3)}\{\mathscr{L}\}$ is improperly orthogonal and according to Eq. (2.86) the group $\Gamma\{\mathscr{H}\}$ (abstract group $\mathscr{H}$) has a normal subgroup $\mathscr{H}^+$ of index 2

$$\mathscr{H} = \mathscr{H}^+ \cup S\mathscr{H}^+ \qquad (3.43)$$

Hence,

$$\Theta^{(\bar{j}j)}_{\bar{\mu}\mu} = h^{-1}\sum_{\mathscr{H}}\Gamma^{(\bar{j})}_{\bar{\mu}'\bar{\mu}}(H)^*\Gamma^{(j)}_{\mu'\mu}(H)\Gamma^{(o-)}(H) =$$

$$= h^{-1}\{\sum_{\mathscr{H}+}\Gamma^{(\bar{j})}_{\bar{\mu}'\bar{\mu}}(H)^*\Gamma^{(j)}_{\mu'\mu}(H) - \sum_{\mathscr{H}+}\Gamma^{(\bar{j})}_{\bar{\mu}'\bar{\mu}}(SH)^*\Gamma^{(j)}_{\mu'\mu}(SH)\} \qquad (3.42b)$$

In the case where

$$S^2 = E, SH = HS \ \ \forall H \in \mathscr{H}^+$$

the irreducible representations of $\mathscr{H}$ occur in pairs $\Gamma^{(j+)}$, $\Gamma^{(j-)}$ of associated representations for which

$$\begin{aligned} &\Gamma^{(jp)}(H) = \Gamma^{(j)}(H), H \in \mathscr{H}^+ \\ &\Gamma^{(jp)}(SH) = (p)\cdot\Gamma^{(j)}(H), H \in \mathscr{H}^+ \end{aligned} \qquad (3.44)$$

Consequently

$$\Theta^{(\bar{j}j)}_{\bar{\mu}\mu} = (2\,l_j^+)^{-1}[1-(\bar{p})(p)]\delta_{\bar{j}j}\delta_{\bar{\mu}'\mu'}\delta_{\bar{\mu}\mu} \qquad (3.42b')$$

hence, in this case a parity selection rule exists

$$\bar{p} \not\longleftrightarrow p, \quad \bar{p} = p$$

$$\bar{p} \longleftrightarrow p, \quad \bar{p} \neq p$$

For a number of typical SRMs the selection rules are given in Tables 14 and 15. The irreducible representations of frequently occurring isometric groups are tabulated in Appendix 5.

2 Magnetic Dipole Transitions

Both WET and selection rules may be derived in a strictly analogous manner by the aid of the transformation formulae (3.36′) and (3.38). One finds

(i) $SO(3)^I$ selection rules: $\Delta J = 0, \pm 1, J = 0 \not\leftrightarrow J = 0$

$\Delta M = 0$ z-polarization

$\Delta M = \pm 1$ x,y-polarization;

(ii) the selection rules w.r.t. $\mathscr{H}(\xi)$ are given the quantity

$$\Phi^{(\bar{J}j)}_{\bar{\mu}\mu} = \frac{1}{h} \sum_{\mathscr{H}} \Gamma^{(\bar{J})}_{\bar{\mu}'\bar{\mu}}(H)^* \Gamma^{(j)}_{\mu'\mu}(H) = l_j^{-1} \delta_{\bar{J}j} \delta_{\bar{\mu}'\mu'} \delta_{\bar{\mu}\mu} \tag{3.45}$$

i.e. for all SRMs, whether belonging to case a or b the magnetic dipole selection rules are

$$\begin{array}{ccc} \Gamma^{(j)} & \leftrightarrow & \Gamma^{(j)} \\ \mu & \leftrightarrow & \mu \end{array}$$

Table 14. Selection rules of dipole transitions of SRMs
f = 1

System/ example	$\mathscr{H}(\xi)$	Dipole selection rules: Electric	Magnetic
$C_sF-C_{3v}T$ CH_3CHO	ϑ_3	$\Gamma^{(o+)} \leftrightarrow \Gamma^{(o-)}$ $\Gamma^{(1)} \leftrightarrow \Gamma^{(1)}$	$\Gamma^{(j)} \leftrightarrow \Gamma^{(j)}$
$C_sF-C_{2v}T$ $CH_2{:}CHNO_2$	$\vartheta_2 \overset{is}{=} \mathscr{V}_4$	$\Gamma^{(1)} \leftrightarrow \Gamma^{(2)}$ $\Gamma^{(3)} \leftrightarrow \Gamma^{(4)}$	$\Gamma^{(j)} \leftrightarrow \Gamma^{(j)}$
C_sF-C_sT $CH_2{:}CHCHO$	$\mathscr{V}_2$	$\Gamma^{(o+)} \leftrightarrow \Gamma^{(o-)}$	$\Gamma^{(j)} \leftrightarrow \Gamma^{(j)}$
$C_sF-C_{2v}I$ $\overline{CH_2CR_2NH}$	$\mathscr{V}_2$	$\Gamma^{(o+)} \leftrightarrow \Gamma^{(o-)}$	$\Gamma^{(j)} \leftrightarrow \Gamma^{(j)}$
$C_{2v}F-C_{2v}I$ $\overline{CH_2CH_2NH}$	$\mathscr{V}_4$	$\Gamma^{(1)} \leftrightarrow \Gamma^{(2)}$ $\Gamma^{(3)} \leftrightarrow \Gamma^{(4)}$	$\Gamma^{(j)} \leftrightarrow \Gamma^{(j)}$
$D_{\infty h}F(C_sT)_2$ CHOCHO	$\mathscr{V}_4$	$\Gamma^{(1)} \leftrightarrow \Gamma^{(2)}$ $\Gamma^{(3)} \leftrightarrow \Gamma^{(4)}$	$\Gamma^{(j)} \leftrightarrow \Gamma^{(j)}$
$C_{2v}F-C_{2v}T$ $C_6H_5NO_2$	$\mathscr{A}(2, 2, 2)$	$\Gamma^{(j+)} \leftrightarrow \Gamma^{(j-)}$	$\Gamma^{(jp)} \leftrightarrow \Gamma^{(jp)}$

Table 15. Selection rules of dipole transitions of SRMs
$\mathscr{G}(\xi) = C_1, f \geq 2$

f	System/ example	$\mathscr{H}(\xi)$	Dipole selection rules: Electric	Dipole selection rules: Magnetic
2	$C_sF(C_{3v}T)(C_{3v}T)'$ $CH_3CH{:}NCH_3$	$\mathscr{G}_{18}$	$\Gamma^{(oo+)} \longleftrightarrow \Gamma^{(oo-)}$ $\Gamma^{(MN)} \longleftrightarrow \Gamma^{(MN)}$	$\Gamma^{(j)} \longleftrightarrow \Gamma^{(j)}$
2	$C_sF(C_{3v}T)(C_{2v}T)$ $CH_3CH_2NO_2$	ϑ_6	$\Gamma^{(o+)} \longleftrightarrow \Gamma^{(o-)}$ $\Gamma^{(3+)} \longleftrightarrow \Gamma^{(3-)}$ $\Gamma^{(1)} \longleftrightarrow \Gamma^{(1)}$ $\Gamma^{(2)} \longleftrightarrow \Gamma^{(2)}$	$\Gamma^{(j)} \longleftrightarrow \Gamma^{(j)}$
2	$C_{2v}F(C_{3v}T)(C_{2v}I)$ CH_3NH_2	ϑ_6	$\Gamma^{(o+)} \longleftrightarrow \Gamma^{(o-)}$ $\Gamma^{(3+)} \longleftrightarrow \Gamma^{(3-)}$ $\Gamma^{(1)} \longleftrightarrow \Gamma^{(1)}$ $\Gamma^{(2)} \longleftrightarrow \Gamma^{(2)}$	$\Gamma^{(j)} \longleftrightarrow \Gamma^{(j)}$
2	$C_sF(C_{2v}T)(C_{2v}T)'$ $NO_2CH{:}CFNO_2$	$\mathscr{A}(2,2,2)$	$\Gamma^{(j+)} \longleftrightarrow \Gamma^{(j-)}$	$\Gamma^{(jp)} \longleftrightarrow \Gamma^{(jp)}$
2	$C_{2v}F(C_{2v}T)_2$ $CH_2(NO_2)_2$	$\vartheta_4\{E, T\}$	$\Gamma^{(o++)} \longleftrightarrow \Gamma^{(o+-)}$ $\Gamma^{(o-+)} \longleftrightarrow \Gamma^{(o--)}$ $\Gamma^{(2++)} \longleftrightarrow \Gamma^{(2+-)}$ $\Gamma^{(2-+)} \longleftrightarrow \Gamma^{(2--)}$ $\Gamma^{(1+)} \longleftrightarrow \Gamma^{(1-)}$	$\Gamma^{(j)} \longleftrightarrow \Gamma^{(j)}$
2	$C_{2v}F(C_sT)_2$ $O(CHO)_2$	$\mathscr{V}_4$	$\Gamma^{(1)} \longleftrightarrow \Gamma^{(2)}$ $\Gamma^{(3)} \longleftrightarrow \Gamma^{(4)}$	$\Gamma^{(j)} \longleftrightarrow \Gamma^{(j)}$
2	$C_sF(C_sT)(C_sT)'$ CH_2FCH_2CHO	$\mathscr{V}_2$	$\Gamma^{(o+)} \longleftrightarrow \Gamma^{(o-)}$	$\Gamma^{(j)} \longleftrightarrow \Gamma^{(j)}$
3	$C_2(\tau)F(C_sT)_2$ CH_2OHCH_2OH	$\mathscr{V}_4$	$\Gamma^{(1)} \longleftrightarrow \Gamma^{(2)}$ $\Gamma^{(3)} \longleftrightarrow \Gamma^{(4)}$	$\Gamma^{(j)} \longleftrightarrow \Gamma^{(j)}$

Again the magnetic dipole selection rules for a number of frequently used SRMs are collected in Tables 14 and 15.

3 Electric Quadrupole Transitions

Starting from Eqs. (3.36″) and (3.38) one obtains for the WET by straightforward calculation

$$\left(\psi_{\bar{J}\bar{M}\Gamma^{(\bar{j})}\bar{\mu}\bar{N}},\ \begin{bmatrix}\hat{A}^1_{00}\\ \hat{A}^1_{2\sigma}\end{bmatrix}\psi_{JM\Gamma^{(j)}\mu N}\right)$$

$$= \sum_{\bar{M}'} \sum_{M'} \sum_{\bar{\mu}'} \sum_{\mu'} (2J+1)^{-1} \begin{bmatrix} \delta_{\bar{J}J}\delta_{\bar{M}'M'}\delta_{\bar{M}M} & 0 \\ & (S_{JM'\bar{J}\bar{M}'2\bar{\sigma}} S_{JM\bar{J}\bar{M}2\sigma}) \end{bmatrix}$$

$$\times \left(\psi_{\bar{J}\bar{M}'\Gamma^{(\bar{J})}\bar{\mu}'N}, \begin{bmatrix} \hat{A}^{I}_{00} \\ \hat{A}^{I}_{2\sigma} \end{bmatrix} \psi_{JM'\Gamma^{(j)}\mu'N} \right) h^{-1} \sum_{\mathscr{H}} \Gamma^{(\bar{J})}_{\bar{\mu}'\bar{\mu}}(H)^* \Gamma^{(j)}_{\mu'\mu}(H) \qquad (3.46)$$

The matrix $(S_{JM'\bar{J}\bar{M}'2\bar{\sigma}} S_{JM\bar{J}\bar{M}2\sigma})$ expresses the electric quadrupole selection rules w.r.t. the group $SO(3)^I$:

$\hat{A}^{I}_{2\sigma}$: $\Delta J = 0, \pm 1, \pm 2; J = 0 \nleftrightarrow J = 0, J = 0 \nleftrightarrow J = 1$

$M = \bar{M} + \sigma, \sigma = -2, -1, 0, +1, +2$

$\hat{A}^{I}_{00}$: $\Delta J = 0, \Delta M = 0.$

The quantity

$$\Xi^{(\bar{J}j)}_{\bar{\mu}\mu} = h^{-1} \sum_{\mathscr{H}} \Gamma^{(\bar{j})}_{\bar{\mu}'\bar{\mu}}(H)^* \Gamma^{(j)}_{\mu'\mu}(H) = l_j^{-1} \delta_{\bar{j}j} \delta_{\bar{\mu}'\mu'} \delta_{\bar{\mu}\mu} \qquad (3.47)$$

again expresses the quadrupole selection rules w.r.t. $\mathscr{H}$ which explicitely read

$\Gamma^{(j)} \leftrightarrow \Gamma^{(j)}$

$\mu \leftrightarrow \mu$

It should be pointed out that the polarizability tensor of a SRM may exhibit a more complicated transformation behavior than expressed by Eq. (3.47). This goes back to the fact, that the polarizability tensor involves all electronic states and the latter do not necessarily all have the same isometric group.

3.4 Chirality of Nonrigid Molecules

The isometric group allows treatment of a number of geometrical problems inherent to SRMs in strict analogy to the group theoretical methods involving the covering symmetry group used for such problems connected with quasirigid molecules. As an important example, the chirality problem of quasirigid molecules should be mentioned, which has been solved most precisely by Kelvin's theorem[42]. According to this theorem, a quasirigid molecule is chiral, if its r_e-structure is improperly congruent with its mirror image. Obviously, it is not possible to apply Kelvin's theorem to nonrigid molecules in a straightforward manner. Nevertheless, it has been applied extensively to nonrigid molecules by considering the covering symmetry of particular NCs of nonrigid molecules, i.e. fixed point configurations in the terminology of this work (cf. Sect. 2.2.3.1). If such NCs with covering operations of the second kind

($|\Gamma^{(3)}(G)| = -1$)[11] exist, the nonrigid molecule has commonly been considered as achiral[43]. Although this criterion leads to correct results in many cases, Mislow has shown that there exist cases where the method contradicts experimental findings, e.g. for molecules of the type

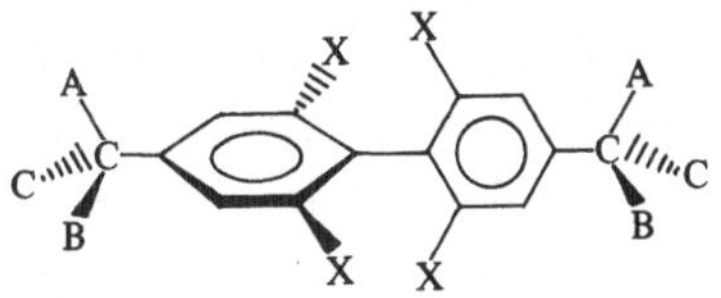

For sufficiently bulky substituents X hindering the internal rotation around the C–C bond between the two phenyl rings and implying a perpendicular conformation of the biphenyl system, there exist no values of the internal rotation angles of the two end-groups CABC leading to a NC with nontrivial covering symmetry. According to the fixed point criterion this molecule should be chiral. This contradicts the experimental finding of Mislow, who found the system to be achiral[44–47].

Basing on the isometric group concept Frei et al.[48] have given a generalization of Kelvins theorem to SRMs and nonrigid molecules (NRM). It may serve as an example for the analogy between the role played by the covering symmetry groups of the r_e-structures of quasirigid molecules and the isometric group of the SRM associated to a NRM, cf. Sect. 4.2.

3.4.1 Theorem for the Chirality of Nonrigid Molecules

First we define: a nonrigid molecule approximated by a SRM with finite internal coordinates will be called chiral, if both conditions (i), (ii) are fullfilled:

(i) no NC in the continuous set of all NC $\{\mathbf{X}_k(\xi), Z_k, M_k\}$ may be mapped onto its mirror image by rotations and translations.

(ii) No NC of this set may be transformed into its mirror image by isometric transformations $\xi' = F(\xi)$, $F \in \mathscr{F}(\xi)$.

This definition is consistent with the definition of the chirality of rigid molecules and forms a sufficient and necessary condition for the optical activity of NRMs. The generalization of Kelvin's theorem for NRMs may be stated as:
a NRM is chiral, if the group $\Gamma^{(3)}\{\mathscr{L}\}$ is properly orthogonal.
The following corollaries hold:

(i) if $\Gamma^{(3)}\{\mathscr{G}\}$ is improperly orthogonal, then every NC $\{\mathbf{X}_k(\xi), Z_k, M_k\}$ is properly congruent with its mirror image and therefore achiral.

(ii) If the group $\Gamma^{(3)}\{\mathscr{K}\}$ is improperly orthogonal and if an isometric transformation F with improperly orthogonal $\Gamma^{(3)}(F)$ has a fixed point, then the NC $\{\mathbf{X}_k(\xi_F), Z_k, M_k\}$ has covering symmetry of the second kind and therefore is achiral.

---

11 I.e., reflections, inversions and rotation-reflections.

The theorem may be supplemented by a few comments:

(i) the theorem shows explicitely the analogy of the roles of the covering symmetry group and the group $\Gamma^{(3)}\{\mathscr{L}\}$ for rigid and nonrigid molecules, respectively.

(ii) The first corollary is a strict analogy for rigid molecules.

(iii) The corollary (ii) is the basis of the conventional procedure for determination of the chirality of NRMs by fixed point nuclear configurations.

(iv) if $\Gamma^{(3)}\{\mathscr{K}\}$ is improperly orthogonal and if it contains improperly orthogonal elements without fixed points the conventional procedure fails. Mislow's molecules are typical examples of this case, as will be demonstrated in the next section.

It should be pointed out that the chirality problem is based entirely on the concept of RNCs. This immediately implies that for its treatment the isometric group $\mathscr{F}(\xi)(\mathscr{H}(\xi))$ is sufficient and the primitive period isometries may be omitted.

3.4.2 Examples

From the examples for construction of isometric groups given in Sect. 2.4, the three SRMs

(i) $D_{\infty h}F(C_1TR)(C_1TS)$
(ii) $D_{2d}F(C_1TR)(C_1TS)$
(iii) $D_{2d}F(C_1TR)_2$
will be used for illustration of the theorem given above.

(i) SRMs of type $D_{\infty h}F(C_1TR)(C_1TS)$: the isometric group has been given in Table 7. The group $\Gamma^{(3)}\{\mathscr{K}\}$ may be taken as

$$\Gamma^{(3)}\{\mathscr{K}\} = \{1^{(3)}, \Gamma^{(3)}(S^f_{12}) = \begin{bmatrix} 1 & & \\ & 1 & \\ & & -1 \end{bmatrix}; -\frac{\pi}{2} < \tau \leqslant +\frac{\pi}{2}$$

or $\Gamma^{(3)}\{\mathscr{K}\} = \{1^{(3)}, -1^{(3)}; 0 \leqslant \tau < \pi\}$
i.e. $\Gamma^{(3)}\{\mathscr{K}\} = C_s$ or $\Gamma^{(3)}\{\mathscr{K}\} = C_i$

Hence $\Gamma^{(3)}\{\mathscr{K}\}$ contains an operation of the second kind (center of symmetry or plane of symmetry, respectively), therefore, according to the theorem molecules of this type are achiral. The isometric transformation F has a fixed point

$$\tau_F = 0 \quad \left(-\frac{\pi}{2} \leqslant \tau < +\frac{\pi}{2}\right)$$

$$\tau_F = \frac{\pi}{2} \quad (0 \leqslant \tau < \pi)$$

According to theorem 2.2.3.1 there belong NCs with C_s or C_i covering symmetry to these fixed points. They are shown schematically in Fig. 7. The role of the primitive period isometric transformation F_3 has been discussed in Sect. 2.4 and is illustrated in a suggestive manner by Fig. 2.

(ii) SRMs of the type $D_{2d}F(C_1TR)(C_1TS)$: the isometric group of the more general SRM $D_2(\tau)F(C_1TR)(C_1TS)$ has been discussed in Sect. 2.4. The isometric

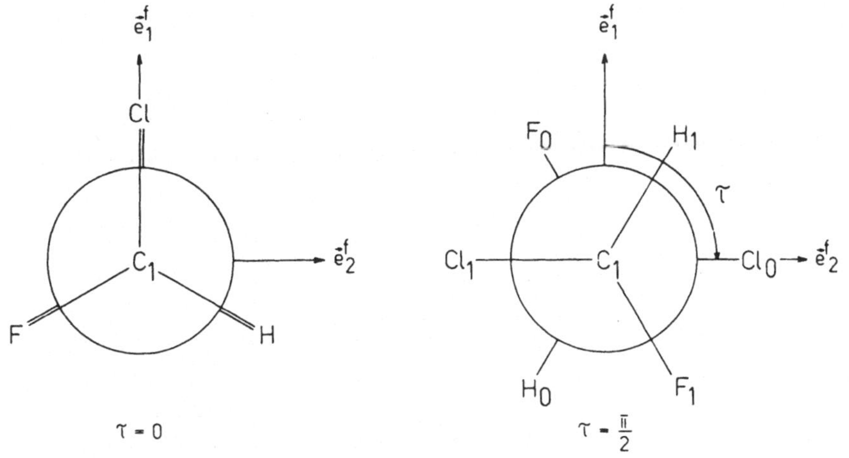

Fig. 7. Fixed point NCs of $D_{\infty h}F(C_1TR)(C_1TS)$ system

Table 16. Isometric group of the $D_{2d}F(C_1TR)(C_1TS)$ system
$\mathscr{G}(v_0, v_1) \overset{\text{is}}{=} \mathscr{C}_1$, $\mathscr{F}(v_0, v_1) \overset{\text{is}}{=} \mathscr{C}_4$

Operator	$\mathscr{A}\{\mathscr{F}\}^a$	$\Gamma^{(NCf)}\{\mathscr{F}\}^b$
E	$\begin{bmatrix} 1 & \cdot & \cdot \\ \cdot & 1 & \cdot \\ \cdot & \cdot & 1 \end{bmatrix}$	$\begin{bmatrix} 1 & \cdot \\ \cdot & 1 \end{bmatrix}$ $\begin{bmatrix} 1 & \cdot & \cdot \\ \cdot & 1 & \cdot \\ \cdot & \cdot & 1 \end{bmatrix}$
F_2	$\begin{bmatrix} \cdot & -1 & \cdot \\ -1 & \cdot & \pi \\ \cdot & \cdot & 1 \end{bmatrix}$	$\begin{bmatrix} \cdot & 1 \\ 1 & \cdot \end{bmatrix}$ $\begin{bmatrix} \cdot & -1 & \cdot \\ 1 & \cdot & \cdot \\ \cdot & \cdot & -1 \end{bmatrix}$
F_3	$\begin{bmatrix} 1 & \cdot & \pi \\ \cdot & 1 & \pi \\ \cdot & \cdot & 1 \end{bmatrix}$	$\begin{bmatrix} 1 & \cdot \\ \cdot & 1 \end{bmatrix}$, $\begin{bmatrix} -1 & \cdot & \cdot \\ \cdot & -1 & \cdot \\ \cdot & \cdot & 1 \end{bmatrix}$
F_4	$\begin{bmatrix} \cdot & -1 & \pi \\ -1 & \cdot & \cdot \\ \cdot & \cdot & 1 \end{bmatrix}$	$\begin{bmatrix} \cdot & 1 \\ 1 & \cdot \end{bmatrix}$, $\begin{bmatrix} \cdot & 1 & \cdot \\ -1 & \cdot & \cdot \\ \cdot & \cdot & -1 \end{bmatrix}$

[a] Representation of $\mathscr{F}$ by substitutions of the internal coordinates:

$$\begin{bmatrix} v_0' \\ v_1' \\ 1 \end{bmatrix} = \begin{bmatrix} A(F) & a(F) \\ 0 & 1 \end{bmatrix} \cdot \begin{bmatrix} v_0 \\ v_1 \\ 1 \end{bmatrix}$$

[b] Representation generated by the vectors $\tilde{X}_{t0}(v_0)$, $\tilde{X}_{t1}(v_1)$ of a set of equivalent nuclei originating from the two equivalent tops (see Fig. 5a).

group of the SRM $D_{2d}F(C_1TR)(C_1TS)$ may be obtained from the group listed in Table 11 by freezing the internal rotation angle τ to the fixed value $\tau = \pi/4$. Physically the freezing process may be achieved by introduction of bulky substituents in the o, o′ positions of the biphenyl system[44, 45]. The resulting isometric group is given in Table 16. From the representation $\Gamma^{(NCf)}\{\mathscr{F}\}$ we conclude

$$\Gamma^{(3)}\{\mathscr{K}\} = S_4 := \left\{ \begin{pmatrix} \cdot & -1 & \cdot \\ 1 & \cdot & \cdot \\ \cdot & \cdot & -1 \end{pmatrix}^k \mid k = 0, 1, 2, 3 \right\}$$

$$\mathscr{A}\{\mathscr{F}\} \overset{\text{is}}{=} \mathscr{C}_4 := \left\{ \begin{pmatrix} \cdot & -1 & \cdot \\ -1 & \cdot & \pi \\ \cdot & \cdot & 1 \end{pmatrix}^k \mid k = 0, 1, 2, 3 \right\}$$

$$\mathscr{G}(v_0, v_1) = C_1$$

hence, since S_4 is improperly orthogonal, the SRM $D_{2d}F(C_1TR)(C_1TS)$ is achiral. However, none of the isometric transformations of $\mathscr{A}\{\mathscr{F}\}$ has a fixed point, hence there exists no NC with second kind covering symmetry. The application of the conventional procedure (search for symmetric NCs) however would predict this SRM to be chiral in contradiction to experimental findings and the chirality theorem 3.4.1.

(iii) SRMs of type $D_{2d}F(C_1TR)_2$: from Table 12 of Sect. 2.4. it follows immediately that

$$\Gamma^{(3)}\{\mathscr{K}\} = D_2$$

hence, SRMs of this type are chiral.

The foregoing discussion shows that the criterion of symmetric NCs does not allow a general decision about the chirality of NRMs. This is the consequence of the fact that isometric transformations do not have fixed points in general. However, the group $\Gamma^{(3)}\{\mathscr{K}\}$ allows a decision about the chirality of SRMs in a simple way, which moreover is strictly analogous to Kelvin's symmetry criterion for quasirigid molecules. One may take this analogy as a further illustration for the fact that the isometric group is a generalization of the conventional symmetry concept of rigid molecules.

3.5 Enumeration and Classification of Conformational Isomers

The isometric group of SRMs has been used for enumeration of the conformational isomers of NRMs[49]. From the point of view of permutational symmetry, this problem has been treated by Mislow et al.[50]. The problem of enumeration of permutational isomers of rigid molecules has been studied by Polya[51] and more generally by Ruch et al.[52]. The determination of classes and number of permutational isomers of molecules with a nonrigid skeleton has been attacked by Leonard[53, 54].

Before presenting an enumeration method based on the isometric group of SRMs the more important assumptions underlying the method should be mentioned:

(i) the method is first formulated for isotopic substitutions. Within the framework of the Born-Oppenheimer approximation all isotopic modifications of a SRM (NRM) have identical sets of NC $\{\mathbf{X}_k(\xi), Z_k\}$[12] and the same electronic energy function $\epsilon^0(\xi)$.

12 The nuclei valued by the charge number Z only.

(ii) For the sake of enumeration of conformational isomers any conceivable, not necessarily isotopic substitution of a nucleus of a SRM is assumed to leave the energy function essentially unchanged, i.e. all conceivable substitution products are assumed to be describable by the same type of SRM. This amounts to an approximate treatment of all conceivable substitution products of a NRM as isotopic modifications. Within the framework of assumption (i) the number of isotopic isomers may be derived. Therefrom a survey over the set of different rotational or vibrational spectra of the isotopic modifications of a SRM with localized r_e-structures may be produced.

3.5.1 Enumeration Theorem for Conformational Isomers

We first consider the internal isometric group $\mathscr{E}(\xi)$ of a SRM defined by the set of nuclear configurations NC $\{\mathbf{X}_k(\xi), Z_k\}$. The isometric group of an isotopic modification of this SRM defined by the set NC $\{\mathbf{X}_k(\xi), Z_k, M_k\}$ is $\mathscr{F}(\xi)$,

$$\mathscr{F}(\xi) \subseteq \mathscr{E}(\xi) \qquad (3.48)^{13}$$

The maximum number of isometric r_e-structures NC $\{\mathbf{X}_k(\xi_e), Z_k\}$ equals $|\mathscr{E}(\xi)|$. This set of r_e-structures decomposes for each particular isotopic modification NC $\{\mathbf{X}_k(\xi), Z_k, M_k\}$ into subsets of isomers (nonisometric NCs). Each isomer is associated with a coset of the decomposition of $\mathscr{E}(\xi)$ modulo $\mathscr{F}(\xi)$ ($e = |\mathscr{E}(\xi)|$, $f = |\mathscr{F}(\xi)|$)

$$\mathscr{E}(\xi) = \bigcup_{k=1}^{e/f} \mathscr{F}(\xi) \cdot E_k \qquad (3.49)$$

the elements of a coset $\mathscr{F}(\xi) \cdot E_k$ representing isometric NC $\{\mathbf{X}_k(\xi_e), Z_k, M_k\}$.

The number of isometric r_e-structures NC $\{\mathbf{X}_k(\xi_e), Z_k\}$ may be a submultiple of $|\mathscr{E}(\xi)|$, namely if the set of internal coordinates of the r_e-structure of NC $\{\mathbf{X}_k(\xi), Z_k\}$ is a fixed point ξ_E of $\xi' = E_1(\xi)$, $\xi' = E_2(\xi)$, ... In this case the number of isometric r_e-structures NC $\{\mathbf{X}(\xi_E), Z_k\}$ is $|\mathscr{E}(\xi)|/|\mathscr{C}(\xi)|$ where $|\mathscr{C}(\xi)|$ is the order of the group $\mathscr{C}$ generated by the periods of $\xi' = E_1(\xi)$, $\xi' = E_2(\xi)$, ...

The set

$$\mathbb{M}_1 := \left\{ \text{NC}\{\mathbf{X}_k(\xi_e'), Z_k, M_k\} \,\middle|\, \begin{pmatrix} \xi_e' \\ 1 \end{pmatrix} = \mathscr{A}(F_k) \cdot \mathscr{A}(C_i) \begin{pmatrix} \xi_e \\ 1 \end{pmatrix},\ F_k \in \mathscr{F}(\xi), C_i \in \mathscr{C}(\xi) \right\} \qquad (3.50)$$

is a set of isometric r_e-structures, since by definition

$$\mathscr{A}(C_i) \begin{pmatrix} \xi_e \\ 1 \end{pmatrix} = \begin{pmatrix} \xi_e \\ 1 \end{pmatrix}$$

13 Since the problem of conformational isomerism is entirely a question of relative nuclear configuration, primitive period isometric transformations have to be omitted.

i.e. all elements of the complex $\mathscr{F}(\xi) \cdot \mathscr{C}(\xi)$ produce the set $\mathbb{M}_1$. $\mathscr{F}(\xi) \cdot \mathscr{C}(\xi)$ is in general not a group. To each double coset of the decomposition of $\mathscr{E}(\xi)$ mod $(\mathscr{F}(\xi), \mathscr{C}(\xi))$

$$\mathscr{E}(\xi) = \bigcup_{i=1}^{n} \mathscr{F}(\xi) E_i \mathscr{C}(\xi) \tag{3.51}$$

represented by the element $E_i \in \mathscr{E}(\xi)$ there corresponds the set of NCs

$$\mathbb{M}_i := \left\{ NC\{\mathbf{X}_k(\xi_e'), Z_k, M_k\} \middle| \begin{pmatrix} \xi_e' \\ 1 \end{pmatrix} = \mathscr{A}(F_k) \cdot \mathscr{A}(E_i) \cdot \mathscr{A}(C_l) \begin{pmatrix} \xi_e \\ 1 \end{pmatrix} \right. \\ \left. F_k \in \mathscr{F}(\xi), C_l \in \mathscr{C}(\xi) \right\} \tag{3.52}$$

which all are isometric among themselves and isomeric to the NC $\in \mathbb{M}_j$, $j \neq i$. Therefore the number of different isomeric sets $\mathbb{M}_i$ of NC $\{\mathbf{X}_k(\xi_e), Z_k, M_k\}$ equals the number of double cosets of the decomposition (3.51). Ruch et al.[52] have given for this number the formula

$$n = \frac{|\mathscr{E}|}{|\mathscr{F}| \cdot |\mathscr{C}|} \sum_{r=1}^{s} \frac{|\mathscr{F} \cap \mathscr{C}_r| \, |\mathscr{C} \cap \mathscr{C}_r|}{|\mathscr{C}_r|} \tag{3.53}$$

where $\mathscr{C}_r$ is the rth class of conjugate elements of $\mathscr{E}$ and the sum is over all classes of $\mathscr{E}$. Since the NC $\{\mathbf{X}_k(F^{-1}(\xi)), Z_k, M_k\}$, $\forall F \in \mathscr{F}(\xi)$ are either properly or improperly congruent n is the number of diastereomers. If we consider the subgroup $\mathscr{F}^+(\xi) \subset \mathscr{F}(\xi)$,

$$\mathscr{F}^+(\xi) := \{F \mid F \in \mathscr{F}(\xi), |\Gamma^{(3)}(F)| = +1\} \tag{3.54}$$

the NC $\{\mathbf{X}_k(F^{-1}(\xi)), Z_k, M_k\}$, $\forall F \in \mathscr{F}^+(\xi)$ are all properly congruent and the number n^+ of double cosets in the decomposition

$$\mathscr{E} = \bigcup_{k=1}^{n^+} \mathscr{F}^+ \cdot E_k \cdot \mathscr{C} \tag{3.51'}$$

is the number of stereoisomeric NCs. Therefore, for SRMs with improperly orthogonal group $\Gamma^{(3)}\{\mathscr{K}\}$ $n^+ - n$ is equal to the number of enantiometric pairs of NCs and $2n - n^+$ equals the number of achiral isomers.

3.5.2 Example

The enumeration method outlined above will be illustrated by determining the number of isomers of various isotopic modifications of diphenylmethane $CH_2(C_6H_5)_2$, assuming various equilibrium values for the internal coordinates. The normal isotopic modification of this nonrigid molecule, and at the same time NC $\{\mathbf{X}_k(\tau_0, \tau_1), Z_k\}$, may be approximated by the SRM $C_{2v}F(C_{2v}T)_2$; another molecule of this type is dini-

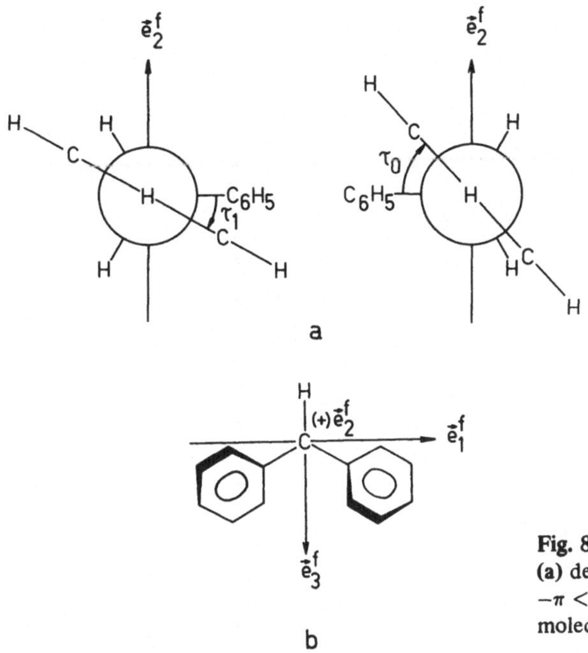

Fig. 8. $C_{2v}F(C_{2v}T)_2SRM$
(a) definition of the internal coordinates $-\pi < \tau_0, \tau_1 \leqslant +\pi$. (b) definition of the molecule fixed coordinate system

tromethane, $CH_2(NO_2)_2$. Figure 8 depicts the choice of the internal coordinates and the frame fixed coordinate system. The internal isometric group $\mathscr{F}(\tau_0, \tau_1) \stackrel{is}{=} \mathscr{A}\{\mathscr{F}\}$ with

$$\begin{bmatrix} \tau_0' \\ \tau_1' \\ 1 \end{bmatrix} = \mathscr{A}(F) \begin{bmatrix} \tau_0 \\ \tau_1 \\ 1 \end{bmatrix}, \quad F \in \mathscr{F}$$

possesses three generators

$$C: \begin{bmatrix} . & 1 & . \\ 1 & . & \pi \\ . & . & 1 \end{bmatrix}, \quad S: \begin{bmatrix} . & 1 & \pi \\ 1 & . & \pi \\ . & . & 1 \end{bmatrix}, \quad T: \begin{bmatrix} -1 & . & \pi \\ . & -1 & \pi \\ . & . & 1 \end{bmatrix}$$

with the relations

$$C^4 = S^2 = T^2 = E, \; SC^k = C^{-k}S, \; TS = ST, \; TC = CT$$

Thus, $\mathscr{F}(\tau_0, \tau_1)$ of the normal isotope and therefore the group $\mathscr{E}(\tau_0, \tau_1)$ is isomorphic to the point group D_{4h}

$$\mathscr{E}(\tau_0, \tau_1) \stackrel{is}{=} (\{C^k\} \cup S\{C^k\}) \cdot (E, T) \stackrel{is}{=} \vartheta_4 \cdot (E, T)$$

A NC with arbitrary τ_0 and τ_1 does not possess covering symmetry, therefore

$$\mathscr{G}(\tau_0, \tau_1) = C_1$$

The representation $\Gamma^{(3)}\{\mathscr{K}\}$ which immediately gives the covering symmetries of the fixed point NCs $NC\{X_k(\tau_{0F}, \tau_{1F}), Z_k\}$ is the point group C_{2v}

Table 17. Dependence of the number of diastereomers and stereoisomers of various isotopic modifications of diphenylmethane on the equilibrium values of the internal coordinates

Isotope/System	$\mathscr{F}(\tau_0, \tau_1)$	τ_{0e}	τ_{1e}	$\mathscr{G}(\tau_{0e}, \tau_{1e})^a$	$\mathscr{C}$	n	n^+
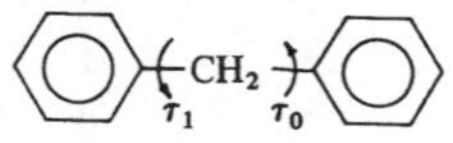 $C_{2v}F(C_{2v}T)_2$	$\vartheta_4 \cdot \{E, T\}$	Arb.	Arb.	C_1	$\mathscr{C}_1$	1	2
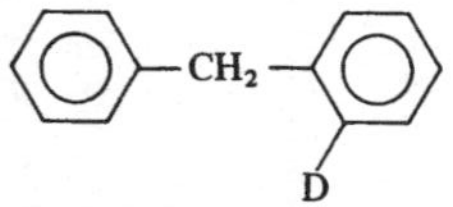 $C_{2v}F(C_{2v}T)(C_sT)$	$\mathscr{V}_4 \overset{is}{=} \{E, SC, TC^2, TSC^3\}$	Arb.	Arb.	C_1	$\mathscr{C}_1$	4	8
		τ_0	τ_0	$C_2{}^b$	$\mathscr{C}_2 \overset{is}{=} \{E, SC^2\}$	2	4
		τ_0	$-\tau_0$	$C_s{}^c$	$\mathscr{C}_2 \overset{is}{=} \{E, TS\}$	2	4
		0	$\pi/2$	$C_s{}^d$	$\mathscr{C}_2 \overset{is}{=} \{E, TSC^3\}$	3	4
		$\pi/2$	$\pi/2$	$C_{2v}{}^e$	$\mathscr{V}_4 \overset{is}{=} \{E, T, SC^2, TSC^2\}$	1	2
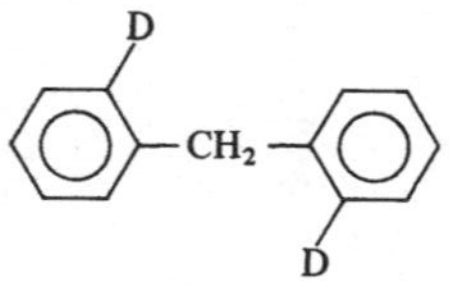 $C_{2v}F(C_sT)_2$	$\mathscr{V}_4 \overset{is}{=} \{E, SC^2, TC^2, TS\}$	Arb.	Arb.	C_1	$\mathscr{C}_1$	4	8
		τ_0	τ_0	C_2	$\mathscr{C}_2 \overset{is}{=} \{E, SC^2\}$	3	6
		τ_0	$-\tau_0$	C_s	$\mathscr{C}_2 \overset{is}{=} \{E, TS\}$	3	4
		0	$\pi/2$	C_s	$\mathscr{C}_2 \overset{is}{=} \{E, TSC^3\}$	2	4
		$\pi/2$	$\pi/2$	C_{2v}	$\mathscr{V}_4 \overset{is}{=} \{E, T, SC^2, TSC^2\}$	2	3
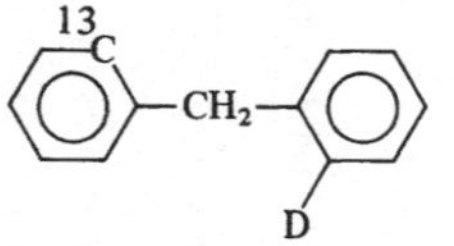 $C_{2v}F(C_sT)(C_sT)'$	$\mathscr{C}_2 \overset{is}{=} \{E, TC^2\}$	Arb.	Arb.	C_1	$\mathscr{C}_1$	8	16
		τ_0	τ_0	C_2	$\mathscr{C}_2 \overset{is}{=} \{E, SC^2\}$	4	8
		τ_0	$-\tau_0$	C_s	$\mathscr{C}_2 \overset{is}{=} \{E, TS\}$	4	8
		0	$\pi/2$	C_s	$\mathscr{C}_2 \overset{is}{=} \{E, TSC^3\}$	4	8
		$\pi/2$	$\pi/2$	C_{2v}	$\mathscr{V}_4 \overset{is}{=} \{E, T, SC^2, TSC^2\}$	2	4

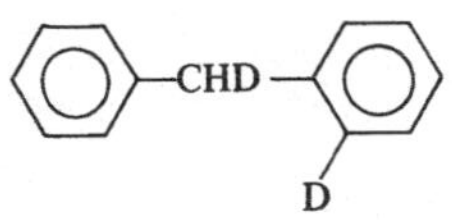	$\mathscr{C}_2 \overset{is}{=} \{E, SC\}$	Arb.	Arb.	C_1	$\mathscr{C}_1$	8	8
		τ_0	τ_0	C_2	$\mathscr{C}_2 \overset{is}{=} \{E, SC^2\}$	4	4
		τ_0	$-\tau_0$	C_s	$\mathscr{C}_2 \overset{is}{=} \{E, TS\}$	4	4
		0	$\pi/2$	C_s	$\mathscr{C}_2 \overset{is}{=} \{E, TSC^3\}$	4	4
$C_sF(C_{2v}T)\,(C_sT)$		$\pi/2$	$\pi/2$	C_{2v}	$\mathscr{V}_4 \overset{is}{=} \{E, T, SC^2, TSC^2\}$	2	2

a Covering group of NC $\{X_k(\tau_{0e}, \tau_{1e}), Z_k\}$.
b $C_2 \overset{is}{=} \{\Gamma^{(3)}(E), \Gamma^{(3)}(C_2(e_3^f))\}$.
c $C_s \overset{is}{=} \{\Gamma^{(3)}(E), \Gamma^{(3)}(S_{23}^f)\}$.
d $C_s \overset{is}{=} \{\Gamma^{(3)}(E), \Gamma^{(3)}(S_{13}^f)\}$.
e $C_{2v} \overset{is}{=} \{\Gamma^{(3)}(E), \Gamma^{(3)}(C_2(e_3^f)), \Gamma^{(3)}(S_{13}^f), \Gamma^{(3)}(S_{23}^f)\}$.

$$\Gamma^{(3)}\{\mathscr{K}\} := \left\{ \begin{pmatrix} 1 & & \\ & 1 & \\ & & 1 \end{pmatrix}, \begin{pmatrix} -1 & & \\ & -1 & \\ & & 1 \end{pmatrix}, \begin{pmatrix} 1 & & \\ & -1 & \\ & & 1 \end{pmatrix}, \begin{pmatrix} -1 & & \\ & 1 & \\ & & 1 \end{pmatrix} \right\} = C_{2v}$$

The SRMs of isotopic modifications of diphenylmethane. e.g. $C_{2v}F(C_{2v}T)(C_sT)$, $C_{2v}F(C_sT)_2$, $C_{2v}F(C_sT)(C_sT)'$, $C_sF(C_{2v}T)(C_sT)$ are all descendents of the SRM $C_{2v}F(C_{2v}T)_2$ of the normal isotope featuring the highest symmetry, and their internal isometric group $\mathscr{F}(\tau_0, \tau_1)$ is a subgroup of $\mathscr{P}(\tau_0, \tau_1)$. The number of isomers of each isotope depends not only on $\mathscr{F}(\tau_0, \tau_1)$ but also on the r_e-values τ_{0e}, τ_{1e} of the internal coordinates, i.e. on the covering group $\mathscr{G}(\tau_{0e}, \tau_{1e})$ of the unsubstituted r_e-structure NC $\{X_k(\tau_{0e}, \tau_{1e}), Z_k\}$. Table 17 shows this dependence of the number of diasteromers (n) and stereoisomers (n^+) for various isotopic modifications.

4 Discussion

In this chapter a few remarks will be stated and relations and extensions of the isometric group concept of semirigid models will be discussed in the following sequence

(i) relation of the isometric group concept to the familiar symmetry concept of quasirigid molecules

(ii) relation of the isometric group of a semirigid model to the isometric group of the associated nonrigid molecule

(iii) remarks concerning the definition of operators $\hat{P}_G$ for semirigid, quasirigid and nonrigid molecules

(iv) remarks concerning the relation of the isometric group to other approaches to the symmetry of nonrigid molecules.

4.1 Isometric Transformations Associated with Quasirigid Molecules

In this section we will show that our group theoretical treatment for nonrigid molecules, approximated by a SRM is strictly analogous to the familiar symmetry treatment of quasirigid molecules (QRM)[55-57]. A QRM may be characterized by the nuclear configurations

$$NC\{X_{ke} + x_k, Z_k, M_k\} \tag{4.1}$$

where the vectors x_k denote infinitesimal displacements of the nuclei from their equilibrium positions X_{ke}. "Infinitesimal" means

$$|d_{kk'} - d_{kk'e}| \ll d_{kk'e} \tag{4.2}$$

Assume the equilibrium nuclear configuration

$$NC_e\{X_{ke}, Z_k, M_k\} \tag{4.3}$$

to have covering symmetry $\mathcal{G}_e$; then, for arbitrary values of the displacements $x_k (k \in [1, K], K \geqslant 4)$, NC (4.1) does not possess covering symmetry

$$\mathcal{G}(x) = C_1 \tag{4.4}$$

whereas

$$\mathcal{G}(0) = \mathcal{G}_e,$$

i.e. the NC with $x_1 = x_2 = \ldots = x_K = 0$ defines a fixed point NC with covering symmetry $\mathcal{G}_e$ in the sense as defined for SRMs, cf. Sect. 2.2.3.1. Though Eq. (4.4) shows that w.r.t. NC (4.1) we are exclusively concerned with internal isometric transformations they will nevertheless be denoted by G since they are a direct consequence of $\mathcal{G}_e$.

To the coordinate transformations

$$(x') = \Gamma^{(3)}(G)(x) \; (\tilde{e}^{f'} = \tilde{e}^{f} \cdot \tilde{\Gamma}^{(3)}(G)) \tag{4.5}$$

expressing the covering symmetry of NC_e are associated operators $\hat{P}_G$ acting in the function space

$$\mathbb{L}\{f(x); f(x) = f(x_1, \ldots, x_{3K})$$
$$\hat{P}_G f(x) = f(\Gamma^{(NCf)}(G)^{-1}(x)) \tag{4.6}$$

In particular, for the set of all displacements x_k arranged in a row

$$\begin{aligned}\hat{P}_G\{\tilde{x}_k\} &= \{\tilde{x}_{\hat{G}k} \cdot \Gamma^{(3)}(G)\} = \{\tilde{x}_k\}\, \Pi(G) \otimes \Gamma^{(3)}(G) \\ &= \{\tilde{x}_k\} \cdot \Gamma^{(NCf)}(G)\end{aligned} \tag{4.7}$$

where $\Pi(G)$ denotes a K by K permutation matrix. This matrix may be specified by requiring that the distance $\hat{P}_G d_{kk'}(x)$ is identical with one of the distances of the set $K(d_{kk'}(x))$, say $d_{\bar{k}\bar{k}'}(x)$.

$$\begin{aligned}\hat{P}_G d^2_{kk'}(x) &= \hat{P}_G((\widetilde{X_{ke} + x_k - X_{k'e} - x_{k'}})(X_{ke} + x_k - X_{k'e} - x_{k'})) \\ &= \hat{P}_G((\widetilde{X_{ke} - X_{k'e}})(X_{ke} - X_{k'e}) + 2(\widetilde{x_k - x_{k'}})(X_{ke} - X_{k'e}) + \\ &\qquad + (\widetilde{x_k - x_{k'}})(x_k - x_{k'})) \\ &= (\widetilde{X_{ke} - X_{k'e}})(X_{ke} - X_{k'e}) + 2(\widetilde{x_{\hat{G}k} - x_{\hat{G}k'}})\,\Gamma^{(3)}(G)(X_{ke} - X_{k'e}) \\ &\quad + (\widetilde{x_{\hat{G}k} - x_{\hat{G}k'}})\,\Gamma^{(3)}(G)\,\tilde{\Gamma}^{(3)}(G)(x_{\hat{G}k} - x_{\hat{G}k'}) \\ &= d^2_{\bar{k}\bar{k}'}(x) = (\widetilde{X_{\bar{k}e} - X_{\bar{k}'e}})(X_{\bar{k}e} - X_{\bar{k}'e}) + 2(\widetilde{x_{\bar{k}} - x_{\bar{k}'}})(X_{\bar{k}e} - X_{\bar{k}'e}) + \\ &\qquad + (\widetilde{x_{\bar{k}} - x_{\bar{k}'}})(x_{\bar{k}} - x_{\bar{k}'})\end{aligned} \tag{4.8}$$

Therefore, since all x_k are arbitrary, the equation

$$\Gamma^{(3)}(G)(X_{ke} - X_{k'e}) = (X_{\bar{k}e} - X_{\bar{k}'e}) \tag{4.9}$$

determines the permutation matrix $\Pi(G)$ in Eq. (4.7).

The matrix group

$$\Pi\{\mathcal{G}_e\} := \{\Pi(G) | \forall G \in \mathcal{G}_e\} \overset{\text{is}}{=} \mathcal{G}_e \tag{4.10}$$

is in general an intransitive permutation group isomorphic to $\mathcal{G}_e$, each set of equivalent nuclei contributing a transitive component. Thereby the transitive components generated by sets in general site w.r.t. $\mathcal{G}_e$ are identical with the regular representation of $\mathcal{G}_e$ generated by right translation[16, 21].

The representation

$$\Gamma^{(NCf)}\{\mathcal{G}_e\} := \{\Pi(G) \otimes \Gamma^{(3)}(G) | \forall G \in \mathcal{G}_e\} \tag{4.11}$$

is isomorphic to $\mathcal{G}_e$, since both the permutational factor $\Pi\{\mathcal{G}_e\}$ and the rotational factor $\Gamma^{(3)}\{\mathcal{G}_e\}$ are isomorphic to $\mathcal{G}_e$. Application of the transformation group $\Gamma^{(NCf)}\{\mathcal{G}_e\}$ to the basis $\{X_{ke}\}$ yields, according to the construction principle fixed by Eqs. (4.7)–(4.9)

$$\{\widetilde{X}_{ke}\} = \{\widetilde{X}_{ke}\}\, \Pi(G) \otimes \Gamma^{(3)}(G),\ \forall G \in \mathcal{G}_e \tag{4.12}$$

This equation again shows that the r_e-structure NC_e is a fixed point NC for all isometric transformations (4.7) and is the analogue of Eq. (2.53) for SRMs. However, for difference concerning the definition of operators $\hat{P}_G$ for SRMs and QRMs see Sect. 4.3.

Application of the operators $\hat{P}_G$ to the nuclear coordinate vectors expressed w.r.t. the LS, $X_k^l(\epsilon, x) = X_{ke}^l(\epsilon) + x_k^l(\epsilon)$, leads to an analogue of the permutation-inversion group [cf. Eq. (2.89)]

$$\begin{aligned}\hat{P}_G\{\widetilde{x}_k^l(\epsilon)\} = \hat{P}_G\{\widetilde{x}_k\}\, 1^{(K)} \otimes R(\epsilon) &= \{\widetilde{x}_k^l(\epsilon)\}\, \Pi(G) \otimes |\Gamma^{(3)}(G)|\, 1^{(3)} \\ &= \{\widetilde{x}_k^l(\epsilon)\}\, \Gamma^{(NCI)}(G),\ \forall G \in \mathcal{G}_e\end{aligned} \tag{4.13}$$

This transformation formula formed the starting point for the study of the symmetry of nonrigid molecules by Hougen[6] and later more generally by Longuet-Higgins[7], cf. Sect. 4.4.1.

From the fact that $\Gamma^{(NCf)}\{\mathcal{G}_e\}$ is the isometric group of the NC (4.1) it follows by the same reasoning as for SRMs that $\mathcal{G}_e$ is the symmetry group of the rotation-vibration hamiltonian. Though the representation (4.11) is commonly used in vibrational spectroscopy[55–57] it only seldom has been characterized as a group of isometric transformations[57].

A further point of interest is the transformation law for vector (tensor) operators w.r.t. the laboratory system. For the electric dipole moment one may show by the same arguments as used for the case of SRMs[41] that

$$\hat{P}_G(\hat{M}_k^{(e)f}(x))\,\hat{P}_G^{-1} = \widetilde{\Gamma}^{(3)}(G)(\hat{M}_k^{(e)f}(x)) \tag{4.14}$$

Hence,

$$\begin{aligned}\hat{P}_G(\hat{M}_k^{(e)l})\hat{P}_G^{-1} &= \hat{P}_G\tilde{R}(\epsilon)(\hat{M}_k^{(e)f})\hat{P}_G^{-1}\\ &= \tilde{R}(\epsilon)R(G)\tilde{\Gamma}^{(3)}(G)(\hat{M}_k^{(e)f})\\ &= \tilde{R}(\epsilon)|\Gamma^{(3)}(G)|(\hat{M}_k^{(e)f})\\ &= |\Gamma^{(3)}(G)|(\hat{M}_k^{(e)l})\end{aligned} \tag{4.15}$$

Equation (4.15) is strictly analogous to Eq. (3.37) for SRMs. It plays therefore the same role in formulation of Wigner-Eckart theorems and selection rules for rotating-vibrating molecules as does Eq. (3.37) for SRMs.

4.2 Relation Between the Isometric Groups of a Nonrigid Molecule and Its Associated Semirigid Model

A NRM may be characterized by the NCs

$$\mathrm{NC}\{\mathbf{X}_k(\xi) + \mathbf{x}_k, Z_k, M_k\},\ k \in [1, K] \tag{4.16}$$

Thereby, each coordinate vector is assumed to be the sum of a vector $\mathbf{X}_k(\xi)$ depending on the finite internal coordinates $\xi_1, \ldots, \xi_f$ and an infinitesimal displacement $\mathbf{x}_k$. The fact that a redundant set of coordinates, i.e. f finite coordinates ξ and 3 K displacements x are associated with NC (4.16) does not affect our discussion of the isometries of the NRM. To the NRM (4.16) there is associated a SRM defined by

$$\mathrm{NC}\{\mathbf{X}_k(\xi), Z_k, M_k\} \tag{4.17}$$

ξ denoting the same set of finite internal coordinates as for NC (4.16). The SRM (4.17) possesses the internal isometric group $\mathscr{F}(\xi)$, the covering group $\mathscr{G}(\xi)$ and the full isometric group $\mathscr{H}(\xi) = \mathscr{F}(\xi) \cdot \mathscr{G}(\xi)$. One immediately may state that the NRM (4.16) does not possess nontrivial covering symmetry $\mathscr{G}(\xi, x)$ if $K \geqslant 4$

$$\mathscr{G}(\xi, x) = C_1 \tag{4.18}$$

and

$$\mathscr{G}(\xi, 0) = \mathscr{G}(\xi)$$

i.e. the NC (4.17) ($x_1 = x_2 = \ldots = x_K = 0$) is a fixed point NC with covering symmetry $\mathscr{G}(\xi)$. Equation (4.18) is strictly analogous to Eq. (4.4) for QRMs and illustrates the similarity of the roles which the r_e-structure and the SRM play for the symmetry of QRMs and NRMs, respectively. The last statement suggests a construction of the isometric group of the NRM, whose elements are of type

$$\begin{bmatrix} \xi' \\ x' \\ 1 \end{bmatrix} = \begin{bmatrix} A & 0 & a \\ . & \Pi \otimes \Gamma^{(3)} & 0 \\ . & . & 1 \end{bmatrix} \cdot \begin{bmatrix} \xi \\ x \\ 1 \end{bmatrix} \tag{4.19}$$

by aid of the isometries of the associated SRM:

(i) A NC (4.16) with arbitrary but fixed values of the internal coordinates ξ possesses the isometric transformations

$$\begin{bmatrix} \xi' \\ x' \\ 1 \end{bmatrix} = \begin{bmatrix} 1^{(f)} & 0 & 0 \\ . & \Pi(G) \otimes \Gamma^{(3)}(G) & 0 \\ . & . & 1 \end{bmatrix} \cdot \begin{bmatrix} \xi \\ x \\ 1 \end{bmatrix} \tag{4.20}$$

with $G \in \mathscr{G}(\xi)$, corresponding to Eq. (4.7) for QRMs. Therefore, the set

$$\left\{ \begin{bmatrix} \xi' \\ x' \\ 1 \end{bmatrix} = \begin{bmatrix} 1^{(f)} & 0 & 0 \\ . & \Pi(G) \otimes \Gamma^{(3)}(G) & 0 \\ . & . & 1 \end{bmatrix} \cdot \begin{bmatrix} \xi \\ x \\ 1 \end{bmatrix} \mid \forall G \in \mathscr{G}(\xi) \right\} \stackrel{\text{is}}{=} \mathscr{G}(\xi) \tag{4.21}$$

forms a subgroup of the isometric group of NC (4.16) isomorphic to $\mathscr{G}(\xi)$. The matrices $\Pi(G) \otimes \Gamma^{(3)}(G)$ are identical with those given by Eq. (2.53) for SRMs, cf. Eq. (4.29).

(ii) Next we consider the group of isometric transformations

$$\left\{ \begin{bmatrix} \xi' \\ x' \\ 1 \end{bmatrix} = \begin{bmatrix} A(F) & 0 & a(F) \\ . & \Pi(F) \otimes \Gamma^{(3)}(F) & 0 \\ . & . & 1 \end{bmatrix} \cdot \begin{bmatrix} \xi \\ x \\ 1 \end{bmatrix} \mid \forall F \in \mathscr{F}(\xi) \right\} \tag{4.22}$$

Thereby, $\begin{pmatrix} A(F) & a(F) \\ 0 & 1 \end{pmatrix} \in \mathscr{A}\{\mathscr{F}\}$ Eq. (2.10) and $\Pi(F) \otimes \Gamma^{(3)}(F) \in \Gamma^{(NCf)}\{\mathscr{F}\}$ Eqs. (2.13), (2.80) for SRMs. It is easily shown that all transformations of the set (4.22) are isometries of NC (4.16)

$$\begin{aligned} d^2_{kk'}(\xi, x) &= |X_k(\xi) + x_k - X_{k'}(\xi) - x_{k'}|^2 \\ &= (\widetilde{X_k(\xi) - X_{k'}(\xi)})(X_k(\xi) - X_{k'}(\xi)) \\ &\quad + 2(\widetilde{x_k - x_{k'}})(X_k(\xi) - X_{k'}(\xi)) + (\widetilde{x_k - x_{k'}})(x_k - x_{k'}) \\ &= D^2_{kk'}(\xi) + 2(\widetilde{x_k - x_{k'}})(X_k(\xi) - X_{k'}(\xi)) + (\widetilde{x_k - x_{k'}})(x_k - x_{k'}) \end{aligned}$$

Application of the operators $\hat{P}_F$ associated with transformations of type (4.22) to $d^2_{kk'}(\xi, x)$ gives

$$\begin{aligned} \hat{P}_F\, d^2_{kk'}(\xi, x) &= D^2_{\bar{k}\bar{k}'}(\xi) \\ &\quad + 2(\widetilde{x_{\bar{k}} - x_{\bar{k}'}})\, \Gamma^{(3)}(F)\widetilde{\Gamma}^{(3)}(F)(X_{\bar{k}}(\xi) - X_{\bar{k}'}(\xi)) \\ &\quad + (\widetilde{x_{\bar{k}} - x_{\bar{k}'}})\, \Gamma^{(3)}(F)\widetilde{\Gamma}^{(3)}(F)(x_{\bar{k}} - x_{\bar{k}'}) \\ &= d^2_{\bar{k}\bar{k}'}(\xi, x) \end{aligned}$$

Hence, the set (4.22) forms a group of isometric transformations of the NRM isomorphic to $\mathscr{F}(\xi)$.

(ii) The semidirect product of the two matrix groups (4.21) and (4.22) generates a group $\mathscr{H}(\xi, x)$ of isometric transformations of NC (4.16) isomorphic to $\mathscr{H}(\xi)$ of NC (4.17). There cannot exist an isometric transformation of the NRM with nontrivial $\mathscr{A}(H)$ not contained in $\mathscr{H}(\xi, x)$, because for any isometry (4.19) not contained in $\mathscr{H}(\xi, x)$,

$$\begin{bmatrix} \xi' \\ 1 \end{bmatrix} = \begin{bmatrix} A & a \\ 0 & 1 \end{bmatrix} \cdot \begin{bmatrix} \xi \\ 1 \end{bmatrix} \tag{4.25}$$

would be an isometry of the associated SRM, in contradiction to the assumption. Therefore, we have the important theorem:
The isometric group $\mathscr{H}(\xi, x)$ of a nonrigid molecule is isomorphic to the isometric group $\mathscr{H}(\xi)$ of the associated SRM.

Representations of $\mathscr{H}(\xi, x)$ on various substrates may be generated along the procedures outlined in Chap. 2. The representation of $\mathscr{H}(\xi, x)$ on the nuclear position vectors

$$\Gamma^{(NCf)}\{\mathscr{H}\} := \{\Pi(H) \otimes \Gamma^{(3)}(H) \mid \forall H \in \mathscr{H}(\xi, x)\} \tag{4.26}$$

is identical with the representation $\Gamma^{(NCf)}\{\mathscr{H}\}$ Eq. (2.73) of the associated SRM, i.e. with the set of solutions (2.70) of Eq. (2.69)

$$\Gamma^{(NCf)}\{\mathscr{H}(\xi, x)\} \equiv \Gamma^{(NCf)}\{\mathscr{H}(\xi)\} \tag{4.27}$$

This follows from the isomorphism between $\mathscr{H}(\xi, x)$ and $\mathscr{H}(\xi)$ and the fact that $X_k(\xi)$ and x_k in the sum $X_k(\xi) + x_k$ must experience the same permutation and rotation under an isometric transformation

$$\begin{aligned} \hat{P}_H \{\widetilde{X_k(\xi) + x_k}\} &= \{\widetilde{X_k(\xi) + x_k}\}\Pi(H) \otimes \Gamma^{(3)}(H) \\ &= \{\widetilde{X}_k(\xi)\}\Pi(H) \otimes \Gamma^{(3)}(H) + \{\widetilde{x}_k\}\Pi(H) \otimes \Gamma^{(3)}(H) \\ \forall H \in \mathscr{H}(\xi, x) \end{aligned} \tag{4.28}$$

In particular, for any $H \in \mathscr{G}(\xi)$

$$\begin{aligned} \hat{P}_G \{\widetilde{X_k(\xi) + x_k}\} &= \{\widetilde{X}_k(\xi)\}\Pi(G) \otimes \Gamma^{(3)}(G) + \{\widetilde{x}_k\}\Pi(G) \otimes \Gamma^{(3)}(G) \\ &= \{\widetilde{X}_k(\xi)\} + \{\widetilde{x}_k\}\Pi(G) \otimes \Gamma^{(3)}(G) \end{aligned} \tag{4.29}$$

cf. Eq. (2.53) for SRMs and Eq. (4.7) for QRMs. Equation (4.29) expresses that any NC of a SRM is a fixed point NC w.r.t. all isometric transformations of type (4.21) of the associated NRM. For difference in definition of the operators $\hat{P}_G$ for SRMs and NRMs, cf. Sect. 4.3. Because the representation $\Gamma^{(NCf)}$ of associated NRMs and SRMs are identical [Eq. (4.27)], the representation $\Gamma^{(NCl)}$ of NRM and SRM must be identical too. Therefore, the relation between isometric group and permutation-inversion group $\mathscr{L}\mathscr{H}$ of NRMs is the same as for the associated SRMs

$$\Gamma^{(NCl)}\{\mathscr{H}(\xi, x)\} \stackrel{is}{=} \mathscr{L}\mathscr{H} \stackrel{is}{=} \mathscr{H}(\xi, x) \tag{4.30}$$

if no primitive period isometric transformation occurs in $\mathscr{H}(\xi, x)$, and

$$\Gamma^{(NCl)}\{\mathscr{H}(\xi, x)\} \stackrel{is}{=} \mathscr{L}\mathscr{H} \stackrel{ho}{=} \mathscr{H}(\xi, x) \tag{4.30'}$$

if $\mathscr{H}(\xi, x)$ contains primitive period transformations, cf. Sect. 2.3.3. It may be shown by the same technique as used for SRMs[14] that $\mathscr{H}(\xi, x)$ is a symmetry group of the rotation-finite internal motion-vibration hamiltonian.

To complete the discussion on the symmetry of QRMs and NRMs it is interesting to point out the analogy between the covering group $\mathscr{G}_e$ of the r_e-structure of a QRM and the isometric group $\mathscr{H}(\xi)$ of the SRM associated with a NRM

$$\begin{aligned} &\text{QRM: } \mathscr{H}(x) \stackrel{is}{=} \mathscr{G}_e \\ &\text{NRM: } \mathscr{H}(\xi, x) \stackrel{is}{=} \mathscr{H}(\xi) \end{aligned} \tag{4.31}$$

These relations show that the isometric group of a SRM plays the same role for a NRM as the covering symmetry group of the r_e-structure plays for a QRM.

4.3 Remarks Concerning the Definition of Operators $\hat{P}_G$

A remark should be made concerning the definition of operators $\hat{P}_G$ for semirigid, quasirigid and nonrigid molecules. These operators are associated with coordinate transformations

$$(X') = \Gamma^{(3)}(G)(X) \tag{4.32}$$

the covering group being defined by the set

$$\{\Gamma^{(3)}(G)\} \stackrel{is}{=} \mathscr{G}(\xi) \tag{4.33}$$

The actual definition of the operators $\hat{P}_G$ for SRMs differs from that for QRMs and NRMs. This originates from the fact that these operators are defined by their action in function spaces which are different for SRMs on one hand and QRMs and NRMs on the other hand. For SRMs the basis is $\{\tilde{X}_k(\xi)\}$, on which $\hat{P}_G$ induces a rotation or a permutation[20]

$$\begin{aligned} &\hat{P}_G\{\tilde{X}_k(\xi)\} = \{\tilde{X}_k(\xi)\}\, 1^{(K)} \otimes \Gamma^{(3)}(G) = \{\tilde{X}_k(\xi)\}\Lambda(G) \otimes 1^{(3)} \\ &\text{or } \hat{P}_G\{\tilde{X}_k(\xi)\} = \{\tilde{X}_k(\xi)\}\Pi(G) \otimes 1^{(3)} \end{aligned} \tag{4.34}$$

$$\begin{aligned} &\text{Therefore } \hat{P}_G\hat{P}_G^{-1}\{\tilde{X}_k(\xi)\} = \{\tilde{X}_k(\xi)\} = \{\tilde{X}_k(\xi)\}\Lambda(G^{-1}) \otimes \Gamma^{(3)}(G) \\ &\Lambda(G^{-1}) = \Pi(G) \end{aligned} \tag{4.36}$$

For QRMs and NRMs, the operators $\hat{P}_G$ are defined by their action on the dynamical variables x_k (infinitesimal displacements)

$$\hat{P}_G\{\tilde{x}_k\} = \{\tilde{x}_k\}\Pi(G) \otimes \Gamma^{(3)}(G) \tag{4.37}$$

with the consequence that

$$\text{QRM: } \hat{P}_G\{\tilde{X}_{ke}\} = \{\tilde{X}_{ke}\} = \{\tilde{X}_{ke}\}\Pi(G) \otimes \Gamma^{(3)}(G) \tag{4.38}$$

$$\text{NRM: } \hat{P}_G\{\tilde{X}_k(\xi)\} = \{\tilde{X}_k(\xi)\} = \{\tilde{X}_k(\xi)\}\Pi(G) \otimes \Gamma^{(3)}(G) \tag{4.39}$$

cf. Eqs. (4.12) and (4.29). Equations (4.36) and (4.39) clearly show the difference in the definition of $\hat{P}_G$ for SRMs and NMRs, respectively.

4.4 Relation of the Isometric Group to Other Approaches

4.4.1 Hougen's Approach and the Longuet-Higgins Permutation-Inversion Group

As mentioned in the introduction, a first attempt to create a symmetry concept for nonrigid molecules has been given by Hougen[6, 58], though the approach strictly speaking applies only to quasirigid molecules or more precisely to molecules with one r_e-structure only. The most important result achieved by this approach is, in our view, the demonstration of the fact, that the rotation-vibration-nuclear spin states may be exhaustively classified in terms of the covering symmetry group of the r_e-structure. By consideration of the transformation properties of the displacement vectors x_k expressed w.r.t. the laboratory system, an equation of the form (4.13) has been obtained. It forms a group which may properly be called the permutation-inversion group of QRMs. Hougen's work formed the starting point of the permutation-inversion approach introduced by Longuet-Higgins[7].

The Longuet-Higgins approach has already been discussed in the introduction and a direct relation between the Longuet-Higgins approach and the isometric group concept has been established by the representation $\Gamma^{(NCl)}\{\mathscr{H}\}$ derived in Sects. 2.3 and 4.2. The discussion presented there made it clear that the two groups (Longuet-Higgins group and isometric group) in general are homomorphic. However, a reconstruction of transformation properties of eulerian angles and internal coordinates expressed in the frame system is not uniquely possible from $\Gamma^{(NCl)}\{\mathscr{H}\}$ alone. By the isometric group concept, such reconstructions are rigorously provided through the representations $\Gamma^{(\mathscr{R})}\{\mathscr{H}\}$ and $\Gamma^{(NCf)}\{\mathscr{H}\}$.

4.4.2 Relation to Altmann's Approach

A constructive relation of the isometric group concept to Altmann's approach[8, 9] in the manner given to Longuet-Higgins' approach would be desirable but appears difficult to be established for the following reasons:

(i) Altmann's Schrödinger group corresponds apparently to the covering group $\mathscr{G}(\xi)$ (covering group of a "random structure"), the "Schrödinger Supergroup" may be considered as the analogue of the full isometric group $\mathscr{H}(\xi)$. The group extension from the Schrödinger group to the supergroup involves isodynamic operations. Though the latter should correspond to the internal isometric transformations, mathematical formulation of the correspondence is not possible, since no definition of the isodynamic operations beyond symbolic operations has to our knowledge been given.

(ii) In certain cases Altmann has proposed for a given SRM several isomorphic supergroups. It has earlier been shown that this phenomenon finds a natural explanation as being automorphisms of the isometric group, induced by transformations of the internal coordinates, which interrelate isometric fixed point NCs. The latter correspond to Altmann's "ordered structures".

(iii) Altmann forwarded the theorem, that the supergroup is always a semidirect product, in which the isodynamic group I plays the role of a normal subgroup. The theorem has been questioned by Watson by presentation of a counter example[59]. Whereas the isometric group has been shown to be decomposable as a semidirect product with the covering group $\mathscr{G}$ (ξ) playing the role of the normal subgroup, Altmann's statement would require to prove the invariance of the group $\mathscr{F}$ (ξ). This, however, cannot be proved and there exist examples which would contradict a theorem of this kind, e.g. the ethylene type molecules discussed in Sect. 2.4.4.

4.4.3 Relation to the Direct Method

As has been mentioned in the introduction the first studies of the symmetry of the rotation-large amplitude motion problem of nonrigid molecules were motivated by spectroscopic investigations of molecules with symmetric internal rotors[11–13]. In this approach the symmetry group of a model hamiltonian for the rotation-internal motion problem is directly investigated for the group of substitutions of the dynamical variables leaving the hamiltonian $\hat{\mathrm{H}}$ Eq. (3.10) symmetric[60]. The relation of this direct approach to the isometric group concept springs from the general theorem, that every isometric transformation is a symmetry of the rotation-internal motion hamiltonian[14]. Hence, the isometric group $\mathscr{H}$ (ξ) is a (proper or improper) subgroup of $\mathscr{G}$ $\{\hat{\mathrm{H}}\}$. In Sect. 3.2.1. special cases have been mentioned, where $\mathscr{H}$ (ξ) is a proper subgroup of $\mathscr{G}$ $\{\hat{\mathrm{H}}\}$. At the present time it appears difficult to derive general conditions for $\mathscr{H}(\xi) \subset \mathscr{G}\{\hat{\mathrm{H}}\}$. Application of the direct method to many SRMs with two symmetric internal rotors have recently been discussed by Dreizler[26, 61].

Appendix 1

Rotation Matrix Parametrized by Eulerian Angles

The rotation matrix $\mathrm{D}(\alpha\beta\gamma)$ describing the relation of the laboratory coordinate system $\tilde{\mathbf{e}}^{\mathrm{l}}$ and the frame system $\tilde{\mathbf{e}}^{\mathrm{f}}$ Eq. (2.1.) is defined as follows ($0 \leqslant \alpha < 2\pi$, $0 \leqslant \beta < \pi$, $0 \leqslant \gamma < 2\pi$). The rotation $\mathrm{D}(\alpha 00)$ transforms $\tilde{\mathbf{e}}^{\mathrm{l}}$ to a new system $\tilde{\mathbf{e}}'$ by rotating $\tilde{\mathbf{e}}^{\mathrm{l}}$ around the $\mathrm{e}_3^{\mathrm{l}}$ axis through α (s: sin, c: cos)

$$\tilde{\mathbf{e}}' = \tilde{\mathbf{e}}^{\mathrm{l}}\,\mathrm{D}(\alpha 00), \quad \mathrm{D}(\alpha 00) = \begin{bmatrix} \mathrm{c}\alpha & -\mathrm{s}\alpha & 0 \\ \mathrm{s}\alpha & \mathrm{c}\alpha & 0 \\ 0 & 0 & 1 \end{bmatrix} \tag{A1.1}$$

Coordinate transformation: $X' = R(\alpha 00)X^l, R(\alpha 00) = \tilde{D}(\alpha 00)$ (A1.1′)

$D(0\beta 0)$ transforms $\tilde{e}'$ to a system $\tilde{e}''$ by rotating $\tilde{e}'$ around the axis e_2' through β

$$\tilde{e}'' = \tilde{e}' D(0\beta 0), D(0\beta 0) = \begin{bmatrix} c\beta & 0 & s\beta \\ 0 & 1 & 0 \\ -s\beta & 0 & c\beta \end{bmatrix} \quad (A1.2)$$

Coordinate transformation: $X'' = R(0\beta 0)X', R(0\beta 0) = \tilde{D}(0\beta 0)$ (A1.2′)

$D(00\gamma)$ transforms $\tilde{e}''$ to the system $\tilde{e}^f$ by rotating $\tilde{e}''$ around the axis e_3'' through γ

$$\tilde{e}^f = \tilde{e}'' D(00\gamma), D(00\gamma) = \begin{bmatrix} c\gamma & -s\gamma & 0 \\ s\gamma & c\gamma & 0 \\ 0 & 0 & 1 \end{bmatrix} \quad (A1.3)$$

$$X^f = R(00\gamma)X'', R(00\gamma) = \tilde{D}(00\gamma) \quad (A1.3')$$

Therefore,

$$\tilde{e}^f = \tilde{e}^l D(\alpha 00) D(0\beta 0) D(00\gamma) = \tilde{e}^l D(\alpha\beta\gamma) \quad (A1.4)$$

$$X^f = R(\alpha\beta\gamma) X^l, R(\alpha\beta\gamma) = \tilde{D}(\alpha\beta\gamma) \quad (A1.4')$$

$$D(\alpha\beta\gamma) = \begin{bmatrix} c\alpha c\beta c\gamma - s\alpha s\gamma & -c\alpha c\beta s\gamma - s\alpha c\gamma & c\alpha s\beta \\ s\alpha c\beta c\gamma + c\alpha s\gamma & -s\alpha c\beta s\gamma + c\alpha c\gamma & s\alpha s\beta \\ -s\beta c\gamma & s\beta s\gamma & c\beta \end{bmatrix} \quad (A1.5)$$

Appendix 2

Transformation Formula for Rotation Group Coefficients

To prove Eqs. (2.88) and (2.91)

$$\hat{P}_H R(\epsilon) = \tilde{R}(H) \cdot R(\epsilon), H \in \mathscr{H}(\xi) \quad (A2.1)$$

and the more general formula (3.33″)

$$\hat{P}_H D^{(s+)}(\epsilon) = D^{(s+)}(H)^\dagger \cdot D^{(s+)}(\epsilon) \quad (A2.1')$$

we start again from the relation between laboratory and frame coordinate system

$$\tilde{e}^f = \tilde{e}^l \cdot D(\epsilon) \quad (A2.2)$$

According to Eqs. (2.25) and (2.61) isometric transformations induce transformations of the frame system as follows ($D(H) = \tilde{\Gamma}^{(3)}(H) = D(\epsilon_H)|D(H)|$)

$$\tilde{e}^{f'} = \tilde{e}^{1}\, D(\epsilon)\, D(H) = \tilde{e}^{1}\, D(\epsilon) D(\epsilon_H) |\, D(H) | \qquad \text{(A2.3)}$$
$$= \tilde{e}^{1}\, D(\epsilon'(\epsilon, \epsilon_H)) |\, D(H) |$$

where $D(H) \in O(3)$, $D(\epsilon) \in SO(3)$

To this equation which defines transformations of the eulerian angles

$$\epsilon' = \epsilon'(\epsilon, \epsilon_H) \qquad \text{(A2.4)}$$

is associated the coordinate transformation of a vector $X \in \mathscr{R}_3(R(\epsilon) = \tilde{D}(\epsilon))$

$$(X)^{\tilde{e}^{f'}} = |\, \Gamma^{(3)}(H) | \cdot R(\epsilon_H) R(\epsilon) (X)^{\tilde{e}^{1}}$$
$$= |\, \Gamma^{(3)}(H) | \cdot R(\epsilon'(\epsilon, \epsilon_H))(X)^{\tilde{e}^{1}} \qquad \text{(A2.5)}$$

Since X is an arbitrary vector

$$R(\epsilon'(\epsilon, \epsilon_H)) = R(\epsilon_H) R(\epsilon) \qquad \text{(A2.6)}$$

Denoting the rotation parameters of the product $R(\epsilon_H) \cdot R(\epsilon)$ by $p(\epsilon_H, \epsilon)$ this transformation may be written as

$$\epsilon' = p(\epsilon_H, \epsilon) \qquad \text{(A2.4')}$$

with $R(p(\epsilon_H, \epsilon)) = R(\epsilon_H) \cdot R(\epsilon)$.

To this transformation we associate the operator $\hat{P}_H$ acting on the linear space $\mathbb{L}\{u(\epsilon)\}$ by the usual convention[18)]

$$\hat{P}_F u(p(\epsilon_F, \epsilon)) = u(\epsilon), \qquad \text{(A2.7)}$$
$$\hat{P}_F u(\epsilon) = u(p^{-1}(\epsilon_F, \epsilon)),$$

where $p^{-1}(\epsilon_F, \epsilon)$ denotes the inverse of $p(\epsilon_F, \epsilon)$ w.r.t. ϵ, i.e. the rotation parameters of the matrix $R^{-1}(\epsilon_F) R(\epsilon)$.

Application of this general formalism to the rotation matrix $R(\epsilon) \in SO(3)$ yields

$$\hat{P}_F R\{p(R(\epsilon_F) R(\epsilon))\} = \hat{P}_F\, R(\epsilon_F) R(\epsilon)$$
$$= R(\epsilon_F)\, \hat{P}_F R(\epsilon) = R(\epsilon), \text{ i.e.}$$
$$\hat{P}_F\, R(\epsilon) = R(\epsilon_F)^{-1} R(\epsilon) \qquad \text{(A2.8)}$$

or for the irreducible representation $D^{(1+)}(\epsilon) = TR(\epsilon)T^{\dagger}$ of SO(3) (the matrix T is given explicitely in Sect. 3.3.1)

$$\hat{P}_H\, D^{(1+)}(\epsilon) = \hat{P}_H TR(\epsilon) T^{\dagger} = TR(\epsilon_H)^{-1} R(\epsilon) T^{\dagger}$$

Thus $\hat{P}_H\, D^{(1+)}(\epsilon) = D^{(1+)}(\epsilon_H)^{\dagger}\, D^{(1+)}(\epsilon)$ (A2.8')

More generally for any representation of SO(3) we have the fundamental formula

$$\hat{P}_H D^{(s+)}(\epsilon) = D^{(s+)}(H)^\dagger D^{(s+)}(\epsilon) \qquad (A2.8'')$$

Appendix 3

Determination of $\Gamma^{(NCf)}(F)$ for SRMs with Proper Covering Group $\mathscr{G}(\xi)$

In this appendix determination of the representation $\Gamma^{(NCf)}\{\mathscr{H}\}$ by solving Eq. (2.69)

$$\hat{P}_F\{\widetilde{X}_k(\xi)\} = \{\widetilde{X}_k(F^{-1}(\xi))\} = \{\widetilde{X}_k(\xi)\}\Pi(F) \otimes \Gamma^{(3)}(F)$$

will explicitely be demonstrated for the isometric transformation F_2 of the SRM $D_{\infty h}F(C_{2v}T)_2$ (Sect. 2.4.4.)

$$\hat{P}_{F_2}\widetilde{X}_{\lambda\mu\nu}(\tau) = \widetilde{X}_{\lambda\mu\nu}(-\tau) = \widetilde{X}_{\lambda'\mu'\nu'}(\tau)\,\Gamma^{(3)}(F_2)$$

Explicitely

$$\left\{\frac{r}{2}(001) + \widetilde{X}^t_{000}\begin{bmatrix}(-1)^\mu & & \\ & (-1)^{\mu+\nu} & \\ & & 1\end{bmatrix}\cdot\begin{bmatrix}c(-\tau) & s(-\tau) & 0\\ -s(-\tau) & c(-\tau) & 0\\ 0 & 0 & 1\end{bmatrix}\right\}\begin{bmatrix}1 & & \\ & -1 & \\ & & -1\end{bmatrix}^\lambda =$$

$$= \left\{\frac{r}{2}(001) + \widetilde{X}^t_{000}\begin{bmatrix}(-1)^{\mu'} & & \\ & (-1)^{\mu'+\nu'} & \\ & & 1\end{bmatrix}\begin{bmatrix}c\tau & s\tau & 0\\ -s\tau & c\tau & 0\\ 0 & 0 & 1\end{bmatrix}\right\}\begin{bmatrix}1 & & \\ & -1 & \\ & & -1\end{bmatrix}^{\lambda'}\cdot\Gamma^{(3)}(F_2)$$

Since X^t_{000} is arbitrary, one may conclude

$$\Gamma^{(3)}(F_2) = \begin{bmatrix}1 & & \\ & -1 & \\ & & -1\end{bmatrix}^{\lambda'}\begin{bmatrix}c\tau & -s\tau & 0\\ s\tau & c\tau & 0\\ 0 & 0 & 1\end{bmatrix}\begin{bmatrix}(-1)^{\mu'} & & \\ & (-1)^{\mu'+\nu'} & \\ & & 1\end{bmatrix}\begin{bmatrix}(-1)^\mu & & \\ & (-1)^{\mu+\nu} & \\ & & 1\end{bmatrix}$$

$$\times\begin{bmatrix}c\tau & -s\tau & 0\\ s\tau & c\tau & 0\\ 0 & 0 & 1\end{bmatrix}\begin{bmatrix}1 & & \\ & -1 & \\ & & -1\end{bmatrix}^{\lambda} = \begin{bmatrix}a_{11} & a_{12} & a_{13}\\ a_{21} & a_{22} & a_{23}\\ a_{31} & a_{32} & a_{33}\end{bmatrix}$$

$$a_{11} = (-1)^{\mu'+\mu}\,c^2\tau + (-1)^{\mu'+\mu+\nu'+\nu+1}\,s^2\tau$$
$$a_{12} = (-1)^{\mu'+\mu+\lambda+1}\,s\tau\,c\tau + (-1)^{\mu'+\mu+\nu'+\nu+\lambda+1}\,s\tau\,c\tau$$
$$a_{13} = 0$$
$$a_{21} = (-1)^{\mu'+\mu+\lambda'}\,s\tau\,c\tau + (-1)^{\mu'+\mu+\nu'+\nu+\lambda'}\,c\tau\,s\tau$$

$$a_{22} = (-1)^{\mu'+\mu+\lambda'+\lambda+1}\, s^2\tau + (-1)^{\mu'+\mu+\nu'+\nu+\lambda'+\lambda}\, c^2\tau$$

$$a_{23} = a_{31} = a_{32} = 0$$

$$a_{33} = (-1)^{\lambda+\lambda'}$$

This matrix must be independent of τ, implying $\nu' = \nu + 1$. Therefore,

$$\Gamma^{(3)}(F_2) = \begin{bmatrix} (-1)^{\mu'+\mu} & 0 & 0 \\ 0 & (-1)^{\mu'+\mu+\lambda'+\lambda+1} & 0 \\ 0 & 0 & (-1)^{\lambda'+\lambda} \end{bmatrix}$$

Furthermore, it must be independent of $\lambda, \lambda', \mu, \mu'$. This requirement leads to the following 4 solutions:

$$\nu' = \nu + 1,\ \lambda' = \lambda,\ \mu' = \mu : (\delta_{\lambda'\lambda}\delta_{\mu'\mu}\delta_{\nu'\nu+1}) \otimes \begin{bmatrix} 1 & & \\ & -1 & \\ & & 1 \end{bmatrix}$$

$$\nu' = \nu + 1,\ \lambda' = \lambda,\ \mu' = \mu + 1 : (\delta_{\lambda'\lambda}\delta_{\mu'\mu+1}\delta_{\nu'\nu+1}) \otimes \begin{bmatrix} -1 & & \\ & 1 & \\ & & 1 \end{bmatrix}$$

$$\nu' = \nu + 1,\ \lambda' = \lambda + 1,\ \mu' = \mu : (\delta_{\lambda'\lambda+1}\delta_{\mu'\mu}\delta_{\nu'\nu+1}) \otimes \begin{bmatrix} 1 & & \\ & 1 & \\ & & -1 \end{bmatrix}$$

$$\nu' = \nu + 1,\ \lambda' = \lambda + 1,\ \mu' = \mu + 1 : (\delta_{\lambda'\lambda+1}\delta_{\mu'\mu+1}\delta_{\nu'\nu+1}) \otimes \begin{bmatrix} -1 & & \\ & -1 & \\ & & -1 \end{bmatrix}$$

The first factor of the direct products denotes a 8 by 8 permutation matrix. If the position vectors are arranged in the order

$$\{\widetilde{X}_{\lambda\mu\nu}\} = \{\widetilde{X}_{000}\widetilde{X}_{001}\widetilde{X}_{010}\widetilde{X}_{011}\widetilde{X}_{100}\widetilde{X}_{101}\widetilde{X}_{110}\widetilde{X}_{111}\},$$

it reads, e.g.

$$(\delta_{\lambda'\lambda+1}\delta_{\mu'\mu}\delta_{\nu'\nu+1}) = \begin{bmatrix} \cdot & \cdot & \cdot & \cdot & \cdot & 1 & \cdot & \cdot \\ \cdot & \cdot & \cdot & \cdot & 1 & \cdot & \cdot & \cdot \\ \cdot & \cdot & \cdot & \cdot & \cdot & \cdot & \cdot & 1 \\ \cdot & \cdot & \cdot & \cdot & \cdot & \cdot & 1 & \cdot \\ \cdot & 1 & \cdot & \cdot & \cdot & \cdot & \cdot & \cdot \\ 1 & \cdot & \cdot & \cdot & \cdot & \cdot & \cdot & \cdot \\ \cdot & \cdot & \cdot & 1 & \cdot & \cdot & \cdot & \cdot \\ \cdot & \cdot & 1 & \cdot & \cdot & \cdot & \cdot & \cdot \end{bmatrix}$$

Appendix 4

Relation Between Irreducible Spherical and Cartesian Components of a Symmetric Tensor of Rank 2

The symmetric part of the Kronecker square $R^{\otimes 2}$, $R \in O(3)$ may be calculated by direct reduction of $R^{\otimes 2}$

$$ZR^{\otimes 2}\tilde{Z} = \begin{bmatrix} R^{s\otimes 2} & \cdot \\ \cdot & R^{a\otimes 2} \end{bmatrix} \tag{A5.1}$$

where $R^{s\otimes 2}$ is a 6 by 6 matrix given by

$$R^{s\otimes 2} = \begin{bmatrix} R_{11}^2 & R_{12}^2 & R_{13}^2 & \sqrt{2}\,R_{11}R_{12} & \sqrt{2}\,R_{12}R_{13} & \sqrt{2}\,R_{11}R_{13} \\ R_{21}^2 & R_{22}^2 & R_{23}^2 & \sqrt{2}\,R_{21}R_{22} & \sqrt{2}\,R_{22}R_{23} & \sqrt{2}\,R_{21}R_{23} \\ R_{31}^2 & R_{32}^2 & R_{33}^2 & \sqrt{2}\,R_{31}R_{32} & \sqrt{2}\,R_{32}R_{33} & \sqrt{2}\,R_{31}R_{33} \\ \sqrt{2}\,R_{11}R_{21} & \sqrt{2}\,R_{12}R_{22} & \sqrt{2}\,R_{13}R_{23} & R_{11}R_{22} + R_{12}R_{21} & R_{12}R_{23} + R_{13}R_{22} & R_{11}R_{23} + R_{13}R_{21} \\ \sqrt{2}\,R_{21}R_{31} & \sqrt{2}\,R_{22}R_{32} & \sqrt{2}\,R_{23}R_{33} & R_{21}R_{32} + R_{22}R_{31} & R_{22}R_{33} + R_{23}R_{32} & R_{21}R_{33} + R_{23}R_{31} \\ \sqrt{2}\,R_{11}R_{31} & \sqrt{2}\,R_{12}R_{32} & \sqrt{2}\,R_{13}R_{33} & R_{11}R_{32} + R_{12}R_{31} & R_{12}R_{33} + R_{13}R_{32} & R_{11}R_{33} + R_{13}R_{31} \end{bmatrix} \tag{A5.2}$$

and $R^{a\otimes 2}$ is the antisymmetric Kronecker square (3 by 3) matrix. Thereby the matrix

$$Z = \begin{bmatrix} 1 & \cdot & \cdot & \cdot & \cdot & \cdot & \cdot & \cdot & \cdot \\ \cdot & \cdot & \cdot & \cdot & 1 & \cdot & \cdot & \cdot & \cdot \\ \cdot & \cdot & \cdot & \cdot & \cdot & \cdot & \cdot & \cdot & 1 \\ \cdot & \sqrt{1/2} & \cdot & \sqrt{1/2} & \cdot & \cdot & \cdot & \cdot & \cdot \\ \cdot & \cdot & \cdot & \cdot & \cdot & \sqrt{1/2} & \cdot & \sqrt{1/2} & \cdot \\ \cdot & \cdot & \sqrt{1/2} & \cdot & \cdot & \cdot & \sqrt{1/2} & \cdot & \cdot \\ \cdot & \sqrt{1/2} & \cdot & -\sqrt{1/2} & \cdot & \cdot & \cdot & \cdot & \cdot \\ \cdot & \cdot & \cdot & \cdot & \cdot & \sqrt{1/2} & -\sqrt{1/2} & \cdot & \cdot \\ \cdot & \cdot & -\sqrt{1/2} & \cdot & \cdot & \cdot & \sqrt{1/2} & \cdot & \cdot \end{bmatrix} \tag{A5.3}$$

transforms the cartesian components of a general tensor $(\hat{A}^f_{mn}(\xi))$ of rank 2

$$\tilde{\hat{A}}^f_{mn}(\xi) = (\hat{A}^f_{11}\hat{A}^f_{12}\hat{A}^f_{13}\hat{A}^f_{21}\hat{A}^f_{22}\hat{A}^f_{23}\hat{A}^f_{31}\hat{A}^f_{32}\hat{A}^f_{33})$$

into symmetric (s) and antisymmetric (a) components

$$\begin{bmatrix} \hat{s}^f(\xi) \\ \hat{a}^f(\xi) \end{bmatrix} = Z(\hat{A}^f_{mn}(\xi)) \tag{A5.4}$$

$$\tilde{\hat{s}}^f(\xi) = (\hat{s}^f_{11}\hat{s}^f_{22}\hat{s}^f_{33}\hat{s}^f_{12}\hat{s}^f_{23}\hat{s}^f_{31})$$
$$\tilde{\hat{a}}^f(\xi) = (\hat{a}^f_{12}\hat{a}^f_{23}\hat{a}^f_{31})$$

The symmetric tensor $(\hat{s}^f(\xi))$ is then transformed to irreducible spherical coordinates by

$$\begin{bmatrix} \hat{A}^f_{00}(\xi) \\ \hat{A}^f_{2\sigma}(\xi) \end{bmatrix} = T^{(2s)}(\hat{s}^f(\xi)) \tag{A5.5}$$

with

$$T^{(2s)} = \begin{bmatrix} -\frac{1}{\sqrt{3}} & -\frac{1}{\sqrt{3}} & -\frac{1}{\sqrt{3}} & \cdot & \cdot & \cdot \\ \frac{1}{2} & -\frac{1}{2} & \cdot & \frac{i}{\sqrt{2}} & \cdot & \cdot \\ \cdot & \cdot & \cdot & \cdot & \frac{i}{\sqrt{2}} & \frac{1}{\sqrt{2}} \\ -\frac{1}{\sqrt{6}} & -\frac{1}{\sqrt{6}} & \sqrt{\frac{2}{3}} & \cdot & \cdot & \cdot \\ \cdot & \cdot & \cdot & \cdot & \frac{i}{\sqrt{2}} & -\frac{1}{\sqrt{2}} \\ \frac{1}{2} & -\frac{1}{2} & \cdot & -\frac{i}{\sqrt{2}} & \cdot & \cdot \end{bmatrix} \tag{A5.6}$$

Appendix 5

Irreducible Representations of Frequently Occuring Isometric Groups

In this work the irreducible representations of the isometric groups occuring in Tables 1, 2, 3, 14, 15 are denoted as follows:

$\mathscr{V}_2$	E	V_2
$\Gamma^{(o+)}$	1	1
$\Gamma^{(o-)}$	1	−1

ϑ_2 =	$\mathscr{V}_4$	E	V_2	V_3	V_4
$\Gamma^{(o+)}$	$\Gamma^{(1)}$	1	1	1	1
$\Gamma^{(o-)}$	$\Gamma^{(2)}$	1	1	−1	−1
$\Gamma^{(1+)}$	$\Gamma^{(3)}$	1	−1	1	−1
$\Gamma^{(1-)}$	$\Gamma^{(4)}$	1	−1	−1	1

$$\vartheta_n,\ n = 2\left[\frac{n}{2}\right], \geqslant 4 \text{ (n even)},\ M = 1, 2, \ldots \left[\frac{n}{2}\right] - 1,\ k = 0(1)\,n-1,\ \varphi = 2\,\pi/n$$

ϑ_n	C^k	SC^k
$\Gamma^{(o+)}$	1	1
$\Gamma^{(o-)}$	1	−1
$\Gamma^{(M)}$	$\begin{pmatrix} \cos\varphi kM & -\sin\varphi kM \\ \sin\varphi kM & \cos\varphi kM \end{pmatrix}$	$\begin{pmatrix} \cos\varphi kM & -\sin\varphi kM \\ -\sin\varphi kM & -\cos\varphi kM \end{pmatrix}$
$\Gamma^{\left(\left[\frac{n}{2}\right]+\right)}$	$(-1)^k$	$(-1)^k$
$\Gamma^{\left(\left[\frac{n}{2}\right]-\right)}$	$(-1)^k$	$-(-1)^k$

$$\vartheta_n,\ n = 2\left[\frac{n}{2}\right] + 1 \text{ (n odd)},\ M = 1, 2, \ldots \left[\frac{n}{2}\right],\ k = 0(1)\,n - 1,\ \varphi = 2\,\pi/n$$

ϑ_n	C^k	SC^k
$\Gamma^{(o+)}$	1	1
$\Gamma^{(o-)}$	1	−1
$\Gamma^{(M)}$	$\begin{pmatrix} \cos\varphi kM & -\sin\varphi kM \\ \sin\varphi kM & \cos\varphi kM \end{pmatrix}$	$\begin{pmatrix} \cos\varphi kM & -\sin\varphi kM \\ -\sin\varphi kM & -\cos\varphi kM \end{pmatrix}$

ϑ_n (E, T) Groups

If T commutes with all $G \in \vartheta_n$ then the representations of $\vartheta_n(E, T)$ are obtained of those of ϑ_n by means of the factor group representations

$\Gamma^{(jp)}$	$\Gamma\{\vartheta_n\}$	$\Gamma\{T\vartheta_n\}$
$\Gamma^{(j+)}$	$\Gamma^{(j)}\{\vartheta_n\}$	$\Gamma^{(j)}\{\vartheta_n\}$
$\Gamma^{(j-)}$	$\Gamma^{(j)}\{\vartheta_n\}$	$-\Gamma^{(j)}\{\vartheta_n\}$

Acknowledgements. We wish to thank The Swiss National Science Foundation (Project Nrs. 2.519–0.76, 2.712–0.77), the ETHZ Administration (Project Nr. 12.611/41) and Messrs. Sandoz AG., Basle, for financial support of this work. Furthermore we wish to thank Dr. P. Groner for valuable discussions and Mrs. R. Zollinger for preparing the manuscript.

5 References

1. Born, M., Oppenheimer, R.: Ann. der Physik *84*, 457 (1927)
2. Pasteur, L.: Lectures at the Société Chimique de Paris, January/February 1860
3. LeBel, J. A.: Bull. Soc. Chim. France *22*, 337 (1874)
4. Coxeter, H. S. M.: Geometry. New York: Wiley 1968
5. Wigner, E. P.: Nachr. Ges. Wiss. Göttingen 1930, p. 130
6. Hougen, J. T.: J. Chem. Phys. *37*, 1433 (1962); *39*, 358 (1963)
7. Longuet-Higgins, H. C.: Mol. Phys. *6*, 445 (1963)
8. Altmann, S. L.: Proc. R. Soc. *A298*, 184 (1967)
9. Altmann, S. L.: Mol. Phys. *21*, 587 (1971)
10. Hougen, J. T.: Can. J. Phys. *42*, 1920 (1964); Can. J. Phys. *43*, 935 (1965); Can. J. Phys. *44*, 1169 (1966)
11. Howard, B.: J. Chem. Phys. *5*, 442 (1937)
12. Wilson, E. Br.: J. Chem. Phys. *6*, 740 (1938)
13. Wilson, E. Br., Lin, C. C., Lide, D. R.: J. Chem. Phys. *23*, 136 (1955)
14. Bauder, A., Meyer, R., Günthard, Hs. H.: Mol. Phys. *28*, 1305 (1974)
15. Frei, H., Meyer, R., Bauder, A., Günthard, Hs. H.: Mol. Phys. *32*, 443 (1976); Mol. Phys. *34*, 1198 (1977)
16. Frei, H., Groner, P., Bauder, A., Günthard, Hs. H.: Mol. Phys. *36*, 1469 (1978)
17. Harary, F.: Graphentheorie. München: R. Oldenbourg 1974, p. 168, theorem 14.1
18. Wigner, E. P.: Group theory. New York: Academic Press 1959, p. 105
19. Kurosch, A. G.: Gruppentheorie II. Berlin: Akademie-Verlag 1972, p. 103
20. Frei, H., Groner, P., Bauder, A., Günthard, Hs. H.: loc. cit., Appendix 1
21. Klingenberg, W., Klein, P.: Lineare Algebra und analytische Geometrie (BI Hochschultaschenbücher Bd. 748, 1972), p. 43
22. Van der Waerden, B. L.: Moderne Algebra, Vol. 1. Berlin, Heidelberg, New York: Springer 1950, p. 150
23. Merer, A. J., Watson, J. K. G.: J. Mol. Spectrosc. *47*, 499 (1973)
24. Dellepiane, G., Gussoni, M., Hougen, J. T.: J. Mol. Spectrosc. *47*, 515 (1973)
25. Gut, M., Meyer, R., Bauder, A., Günthard, Hs. H.: Chem. Phys. *31*, 433 (1978)
26. Dreizler, H., in: Fortschr. chem. Forsch. *10*, p. 59 (1968)
27. Bunker, P. R., in: Vibrational spectra and structure, Vol. 3. Durig, J. R. (ed.). New York: M. Dekker, Inc. 1975, p. 1
28. Groner, P., Durig, J. R.: J. Chem. Phys. *66*, 1856 (1977)
29. Wigner, E. P.: loc. cit., p. 325
30. Edmonds, A. R.: Angular momentum in quantum mechanics. Princeton University Press 1957, p. 7. The definitions for the eulerian angles defined in this book will be used throughout this paper
31. Kemble, E. C.: The fundamental principles of quantum mechanics. New York: Dover Publications 1958, p. 237
32. Podolsky, B.: Phys. Rev. *32*, 812 (1928)
33. Wigner, E. P.: loc. cit., p. 112
34. Wigner, E. P.: loc. cit., p. 167, Eq. (15.27)
35. Meyer, R., Günthard, Hs. H.: J. Chem. Phys. *49*, 1510 (1968)
36. Bauder, A., Mathier, E., Meyer, R., Ribeaud, M., Günthard, Hs. H.: Mol. Phys. *15*, 597 (1968)
37. Attanasio, A., Bauder, A., Günthard, Hs. H., Keller, J.: Mol. Phys. *21*, 35 (1971)
38. Baltagi, F., Bauder, A., Henrici, P., Ueda T., Günthard, Hs. H.: Mol. Phys. *24*, 945 (1972)
39. Wigner, E. P.: loc. cit., p. 114, Eq. (12.6)
40. Boerner, H.: Darstellungen von Gruppen. Berlin, Göttingen, Heidelberg: Springer 1955, p. 90
41. Frei. H., Meyer, R., Bauder, A., Günthard, Hs. H.: loc. cit., Appendix 1
42. Kelvin, Lord: Baltimore lectures 1884. In: Baltimore lectures. London: C. J. Clay and Sons 1904, pp. 436, 619

43. Eliel, E. L.: Stereochemistry of carbon compounds. New York: McGraw-Hill Book Comp. 1962, p. 11
44. Mislow, K.: Science *120,* 232 (1954)
45. Mislow, K., Bolstad, R.: J. Am. Chem. Soc. *77,* 6712 (1955)
46. Mislow, K.: Trans. N. Y. Acad. Sci. *19,* 298 (1957)
47. Mislow, K.: Introduction to stereochemistry. New York: Benjamin 1966, p. 93
48. Frei, H., Günthard, Hs. H.: Chem. Phys. *15,* 155 (1976)
49. Frei, H.: Thesis Nr. 5905, ETH Zürich (1977)
50. Mislow, K., Gust, D., Finocchiaro, P., Boettcher, R. J.: Topics Curr. Chem. *47,* 1 (1974)
51. Polya, G.: Acta Math. *68,* 145 (1937)
52. Ruch, E., Hässelbarth, W., Richter, B.: Theoret. Chim. Acta *19,* 288 (1970)
53. Leonard, J. E., Hammond, G. S., Simmons, H. E.: J. Am. Chem. Soc. *97,* 5052 (1975)
54. Leonard, J. E.: J. Phys. Chem. *81,* 2212 (1977)
55. Wilson, E. Br., Decius, J. C., Cross, P. C.: Molecular vibrations. New York: McGraw-Hill Book Company 1955, p. 88
56. Califano, S.: Vibrational states. New York: Wiley 1976, p. 150
57. Günthard, Hs. H., Gäumann, T., Heilbronner, E.: Helv. Chim. Acta *32,* 178 (1949)
58. Hougen, J. T.: J. Chem. Phys. *55,* 1122 (1971)
59. Watson, J. K. G.: Mol. Phys. *21,* 577 (1971)
60. Whittaker, E. T.: Analytical dynamics. New York: Dover Publications 1944, p. 288
61. Dreizler, H., in: Molecular spectroscopy: Modern research. Rao, K. N. R., Mathews, C. W. (eds.). New York: Academic Press 1972, p. 59

Structure of Molecules with Large Amplitude Motion as Determined from Electron-Diffraction Studies in the Gas Phase

Otto Bastiansen, Kari Kveseth, and Harald Møllendal

Department of Chemistry, University of Oslo, Box 1033, Blindern, Oslo 3, Norway

Table of Contents

1 Introduction

During the last fifty years, since the famous experiment of Davisson and Germer[1], electron-diffraction methods have been widely in use for the study of structure of matter on a molecular scale. The use of electron diffraction in the gas phase provided an excellent method for the study of the structure of the free molecule. In the earlier years of gas electron diffraction the concept "molecular structure" was in practice synonymous with molecular geometry. The internal motion of the molecule was almost considered by electron diffractionists as an unavoidable evil, a defect of nature interfering unfortunately with the endeavor to determine molecular geometry. In order to minimize the role of this "defect" it was found unseful to apply spectroscopically obtained data on the internal motion of the molecule. The idea was to correct for the effect of intramolecular motion, and thus try do develop a procedure for obtaining the best geometric data for an idealized rigid molecular model. However, the development of the electron-diffraction method for studies in the gas phase soon changed the ambitions of the researchers using the method. The potentiality of the method proved considerably greater than generally expected in the earlier days of electron diffraction. The development of the method led to considerable improvement in the precision of intensity data measurements. Further dramatically improved computing procedures made it possible to deduce the molecular data latent in the measurements faster and with considerably greater reliability. Thus it was soon evident that data characterizing intramolecular motion could also be quantitatively obtained from electron-diffraction measurements. The concept "molecular structure" in the modern practice of gas electron diffraction is accordingly no longer restricted to geometry alone, but includes parameters describing internal motion as well.

The internal motion of a polyatomic molecule is rather complex. In order to understand its principles, it is convenient and customary to refer to the vibration of a diatomic molecule. The internal motion in molecules has above all been studied experimentally and theoretically by spectroscopists. The approach towards the study of polyatomic molecules through diatomic molecules has been excellently demonstrated by the two classical books of G. Herzberg, the first one dealing with diatomic molecules[2], the second one with polyatomic molecules[3]. Although these two books should by now be hopelessly out of date, and in spite of the fact that they have been succeeded by a series of up to date works, they can still be recommended to to-day's students.

The two key quantities in quantum mechanics are the eigenfunction and the eigenvalue. While the eigenvalue, through energy difference, is the quantity that in principle is attainable by spectroscopic studies, electron-diffraction studies in the gas phase give information about the eigenfunction, or rather the square of the eigenfunction of the atom distribution. Let us take as an example a diatomic molecule. The radial distribution curve as determined by gas electron diffraction, if properly modified, is a description of the temperature average of the distance distribution in the diatomic molecule (Fig. 1).

The position of the peak on a distance scale gives information of the internuclear distance, and the shape of the peak itself is a representation of the weighted sum of

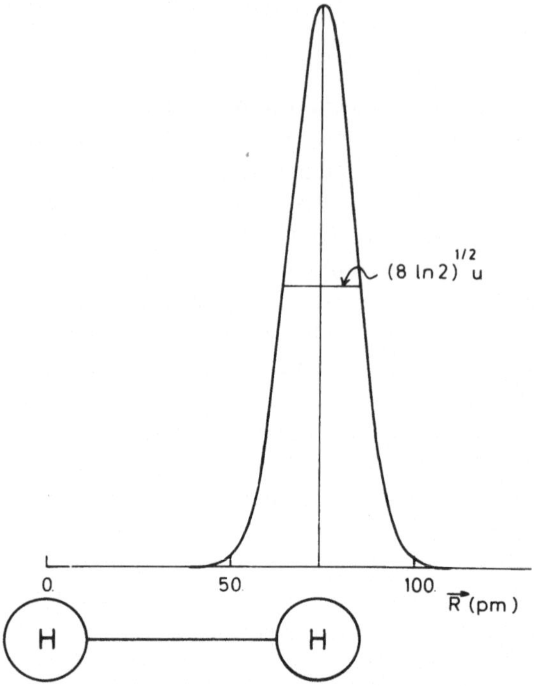

Fig. 1. Radial distribution curve for a diatomic molecule (H_2; R = 74, u = 8,7 (pm))

the $\psi^*\psi$ for the interatomic distance at the temperature of the experiment. The radial distribution curve obtained by electron diffraction for a diatomic molecule should in principle contain the data necessary for describing the molecular vibration, including both the harmonic and the anharmonic contribution. However, the precision by which spectroscopic methods produce such data usually exceeds what may be attained by electron diffraction, at least for small molecules.

In the case of polyatomic molecules the radial distribution curve as deduced from electron-diffraction gas experiments may also be considered as a kind of a weighted sum of $\psi^*\psi$ curves for the internal motion in the molecule, but here all internuclear distances are inseparably mixed together in a one-dimensional representation. For a "rigid" molecule, such as carbon tetrachloride or benzene, electron diffraction may produce quite accurate information as to the geometry of the molecule. As to the internal motion of the molecule, vibrational amplitudes may be deduced and compared with the corresponding data, differently but usually considerably more accurately, obtained by spectroscopic methods. How this is actually done in practice is perhaps most elegantly described by S. Cyvin[4)].

The concept "rigid" molecule is of course, like many other useful terms in science, not well defined. The distinction between a "rigid" and a "non-rigid" molecule is in practice based upon and left to the intuition of the chemists. The same is the case with the key concept of the present article, namely the term "large amplitude motion". It would hardly be practicable to define the difference between "small" and "large" amplitude motion quantitatively by a certain value for the amplitude dividing intramolecular motion into two categories. The term "large amplitude motion" has to be applied to certain kinds of internal molecular motion, the description of which is the aim of the present article. The "small" amplitudes would in practice be of the order of approximately 4 pm up to values ranging from 10 to 20 pm. The lower value of a "large" amplitude may well overlap with the upper limit of a "small" amplitude, but this slight blemish would hardly lead to confusion.

In the electron-diffraction jargon it is often referred to "the framework vibration" in contrast to the large amplitude motion. The idea is to try to separate the large amplitude motion, as for example a torsional motion, from the small amplitude vibration also taking place in "rigid" molecules. This practical approach does not lead to semantic difficulties, but the approach, of course, meets with the well known difficulty in any theoretical treatment of this kind, namely the problem of separability of the energy and consequently of the Hamiltonian operator.

1.1 Intramolecular Motion and Conformational Analysis

The electron-diffraction method presented the first demonstration of the existence of conformational equilibrium in the gas phase[5]. Later a series of examples of conformational equilibrium changes as function of temperature have been described[6–8]. The transformation from one conformer to another is considered to take place without breaking of bonds and is thus a typical example of large amplitude motion within the molecule. But in spite of the fact that gas electron diffraction has been a key method for proving the existence of this kind of large amplitude motion, it unfortunately fails to give accurate description of the mechanism of the motion. The gas electron-diffraction intensity data are composed of contributions from all interatomic distances existing in the molecular species studied, and in such a way that the contribution from each individual distance value is proportional to the probability of observing the said value. The contribution of the equilibrium conformers (usually simply referred to as the conformers) to the electron-diffraction intensity data is often overwhelmingly higher than that of the intermediate geometrical species of the molecule. As an example may serve the classical case of a dihalocyclohexane (see Fig. 17) or a 1,2-dihaloethane. The main indicator for the two coexisting conformers is the halogen-halogen distance. This distance shows up in the radial distribution curve with two peaks, one corresponding to each of the two conformers. Even the most refined modern electron-diffraction method applied on such molecules would only give very limited, if any, measurable contribution from intermediate molecular species deviating appreciably from the two conformers. The electron-diffraction method needs a rather high probability value for a geometric species in order to recognize it. The intramolecular motion between conformers can thus in general not be studied directly by electron diffraction alone except for the motion near by or at the equilibria. Only in cases with small barriers between the conformers, i. e. about or smaller than RT, appreciable amounts of intermediates may be recognizable.

The internal motion of a molecule would be satisfactorily and adequately described if the potential energy of the molecule as function of a complete set of independent geometrical variables had been given. For the intramolecular motion leading from one conformer to another, the lowest energy pass would be of particular interest. The electron-diffraction method may in principle help finding

1) the position of the minima on such an energy curve,
2) the energy difference between the minima, and
3) the shape of the curve in the minimum areas.

In practice some of these data may be hard to obtain from electron diffraction alone with a meaningful accuracy. And in any case, a full description of the energy curve leading from one conformer to another can only be obtained in combination with other methods.

The lowest energy pass leading from one conformer to another may be described by a small number of geometrical parameters. In an ethane derivative, as for example 1,2-dihaloethane, this pass may be described to the first approximation by one parameter, namely the torsional angle, though the actual lowest energy pass probably involves bendings of valency angles and even stretching of bonds. For cyclohexane derivatives the lowest energy pass must be described by at least two parameters, the torsional angle and the C-C-C valency angle. Other parameters would probably also change during a ring conversion. For cyclohexanes the lowest energy pass is assumed to go through a geometric form of a cyclohexene-like half chair.

The kind of reasoning used concerning the transformation from one conformer to another may of course be used more generally than for the examples mentioned exhibiting only two conformers. It may of course also be used in cases such as ethane or cyclohexane itself or in any of their derivatives where large internal motion of the same kind is described. The reasoning may also easily be carried out for more complex molecules involving several torsional parameters and exhibiting more than two conformers.

2 The Electron-Diffraction Method

Both the experimental technique[9)] and theory[10)] behind structure determination using the gas electron-diffraction method have been reviewed in detail by several authors. Since the experiments for the study of large amplitude motion are the same as for other electron-diffraction investigations, it was not felt necessary to describe the experimental equipment and procedure. There is only one experimental point that deserves a comment. For quantitative studies of large amplitude motion the temperature should be known. In particular for conformational analysis temperature studies are important since the relative amount of coexisting conformers is temperature dependent. In extreme cases an existing conformer may even get lost in a conventional low temperature experiment, while it clearly shows up at high temperatures. This was most clearly demonstrated in the case of 2-chloroethanol where only the *gauche* conformer showed up in the original electron-diffraction study based upon optimal temperature exposures[11)], while the *anti* conformer clearly showed up at higher temperatures[7,8)] (Fig. 6).

There are several difficulties assosiated with electron-diffraction temperature experiments. Firstly, there is no direct way to measure the temperature of the gas at the moment of diffraction. Secondly, the concept of temperature is an ill-defined parameter in the diffraction point. The gas to be studied is let out of a nozzle into the vacuum system and immediately hit by the electron beam. The gas is accordingly in the moment of diffraction expanding into the low pressure area. One could therefore hardly characterize the situation as a thermodynamic equilibrium. The only experimental temperature to be referred to is the one of the nozzle tip which temperature is measured by a thermocouple. However, experience, particularly in connection with the measurements of conformational equilibria, seems to indicate that the cooling of the gas caused by the expansion primarily concern the translational motion of the molecules, to a lesser degree the overall rotation, and only to a minor degree the internal motion which is the main interest in present context. Accordingly, in most electron-diffraction work the temperature of the nozzle tip is taken as the temperature of internal motion, though attempts to correct for the cooling of the gas have also been made[12].

As to the theoretical part of electron-diffraction studies it was felt appropriate, in order to ease the presentation of the problem of large amplitude molecular motion, to include some basic equations and to discuss some of the approximations made.

The intensity function used in structure analyses may be expressed as follows:

$$I(s) = \text{const.} \sum_{i \neq j}^{M} \sum^{M} g_{ij/kl}(s) \int P_{ij}(r) \frac{\sin rs}{r} dr \tag{1}$$

where

$$g_{ij/kl}(s) = \frac{|f_i(s)| \cdot |f_j(s)|}{|f_k(s)| \cdot |f_l(s)|} \cos(\eta_i(s) - \eta_j(s)) \tag{2}$$

The independent variable $s = \frac{4\pi}{\lambda} \sin \theta$, where λ is the electron wavelength and 2θ the scattering angle. The summation indices i and j refer to each of the M atoms in the molecule. The index pair k and l refers to a representative atom pair of the molecule studied, chosen to obtain a convenient form for I(s) and its Fourier transformed partner. $f_i(s)$ is the complex scattering amplitude of the i-th atom in the molecule and $\eta_i(s)$ is the argument of $f_i(s)$, *i. e.*

$$f(s) = |f(s)| \exp(i\eta(s))$$

The electron distribution of the atoms are assumed to be spherically symmetric. $P_{ij}(r)\,dr$ is the probability of finding the atom pair i and j at an internuclear distance interval between r and r+dr. The integral over r, although for convenience written without specified limits of integration, should be considered as a definite integral. $P_{ij}(r)$ is always limited to a rather narrow r-interval beyond which it is negligible in value.

Calculations of f- and η-values from atomic potentials are continuously being made more comprehensive and accurate[13] and tabulated values are available[14].

I(s) as given in Eq. (1) is in principal of experimental origin though it is not the directly obtained intensity value. It represents a properly modified difference curve between the total observed intensity curve and a background intensity curve. The shape of the latter curve is independent of the molecular structure and is only reflecting the scattering properties of the atoms in the molecule.

For a homonuclear diatomic molecule Eq. (1) gets the simple form:

$$I(s) = \text{const.} \int_0^\infty \frac{P(r)}{r} \sin rs \, dr \tag{3}$$

Here integration limits are included.

By Fourier transformation $\frac{P(r)}{r}$ may be expressed as

$$\frac{P(r)}{r} = \text{const.} \int_0^\infty I(s) \sin rs \, ds \tag{4}$$

In the general polyatomic case a function is defined in analogy to Eq. (4) as follows:

$$\frac{\sigma(r)}{r} = \int_{s_{min}}^{s_{max}} I(s) \exp(-ks^2) \sin rs \, ds \tag{5}$$

This function is called the radial distribution (RD) function. The integration limits are set by experimental insufficiency. In routine gas electron diffraction $s_{min} > 0.01$ pm^{-1} and $s_{max} < 0.6$ pm^{-1}, though in especially designed experiments these limits may be exceeded. The lack of experimental information on the lower side of the s range may be remedied by introduction of theoretical intensity values. The effect of the outer limit is reduced by the factor $\exp(-ks^2)$ where k is chosen from experience.

For interatomic distances undergoing small vibrational motion, a Gaussian distance distribution represents a good approximation for $P_{ij}(r)$:

$$P_{ij}(r) = \frac{1}{\sqrt{2\pi}\, u_{ij}} \exp\left[-\frac{(r-r_{ij})^2}{2u_{ij}^2}\right] \tag{6}$$

where u_{ij} is the root-mean-square amplitude of vibration and r_{ij} the mean distance between the two atoms involved. For an atom pair where the two atoms undergo a mutually large amplitude motion, a situation in focus of the present article, $P_{ij}(r)$ is more complex.

In order to perform the integration in Eq. (1) or (3), r is set equal to $r_{ij}+y$, y being of a small value compared to r_{ij}. We then have:

$$\frac{1}{r} = \frac{1}{r_{ij}} \left(1 - \frac{y}{r_{ij}} + \ldots\right) \tag{7}$$

Using only the first term of Eq. (7) and the approximation set by Eq. (6) the integral may be written as

$$\int P_{ij}(r) \frac{\sin rs}{r} \, dr = \frac{\sin r_{ij}s}{r_{ij}} \exp\left(-\frac{1}{2} u_{ij}^2 s^2\right) \tag{8}$$

The approximations needed for carrying out the integration are for all purposes sufficiently good.

Equation (1) may thus be written

$$I(s) = \text{const.} \sum_{i \neq j}^{M} \sum^{M} g_{ij/kl}(s) \frac{\sin r_{ij}s}{r_{ij}} \exp\left(-\frac{1}{2} u_{ij}^2 s^2\right) \tag{9}$$

To the approximation applied so far and after correction for the lack of information beyond the experimental s-range, the RD-curve should consist of Gaussian shaped curves for distances between the kind of atoms described by the indices k and l. The maximum of each Gaussian peak corresponds to the mean of the interatomic distance in question. For other pairs of atoms the corresponding peak is slightly modified due to the difference in f(s) and η(s) for the various atoms. But at this level of approximation a series of effects have been neglected. These are:

(1) further terms in Eq. (7),
(2) anharmonicity,
(3) shrinkage effect, and
(4) large amplitude motion.

1) Inclusion of one more term in the series Eq. (7) leads to a change in Eq. (8). Instead of sin (r_{ij}s) one gets to a good approximation $\sin\left(r_{ij} - \frac{u_{ij}^2}{r_{ij}}\right)s$. The distance most directly related to the electron-diffraction study is denoted r_a [10, 15]. The average interatomic distance is

$$r_g = r_a + \frac{u^2}{r} \tag{10}$$

The correction term $\frac{u^2}{r}$ is usually small compared to r_g, ranging from 0.1 to 0.5 pm.

2) Anharmonicity is conveniently treated using the Morse-potential approximation. The sine term of Eq. (8) is then to a first approximation replaced by $\sin\left(r_{ij} - \frac{u_{ij}^2}{r_{ij}} - \kappa_{ij}s^2\right)s$ where κ_{ij}, the asymmetry constant, is related to the constant a in the Morse-potential[16] through the approximation $\kappa = au^4/6$. A more accurate expression for diatomic molecules is

$$\kappa = \frac{au^4}{6}[1 + 8\chi/(1 + \chi^2)] \tag{11}$$

where $\chi = h\nu/kT$.

3) The shrinkage effect[17] is treated in more detail elsewhere in the present article. Due to molecular vibrations interatomic distances observed by electron diffraction do not correspond to a set of distances calculated from a rigid geometrical model. Usually the shrinkage effect is routinely included in electron-diffraction least-squares refinement. In order to do so, it has been found appropriate to introduce a third distance type r_α defined as the distance between mean positions of atoms at a particular temperature. If the harmonic force field is known, r_α may be calculated from r_a according to Eq. (12):

$$r_\alpha = r_a + \frac{u^2}{r} - K \qquad (12)$$

where $K = \dfrac{\langle \Delta x^2 \rangle + \langle \Delta y^2 \rangle}{2r}$,

$\langle \Delta x^2 \rangle$, and $\langle \Delta y^2 \rangle$ are the mean square perpendicular vibrational amplitudes. Equation (12) demonstrates two features: a) even for negligible perpendicular amplitudes shrinkage will be observed in the r_a-value, b) shrinkage also occurs in the harmonic approximation for P(r).

4) For large amplitude motion the distance distribution P(r) may be rather complex. For most molecules studied by electron diffraction, the large amplitude motion may be described by a few geometrical parameters, and in many important cases only by one single parameter. In addition to the large amplitude motion the molecule will of course carry out the same kind of small amplitude vibration as any other molecule. This vibration, which is referred to as the framework vibration, is assumed to be separable from the large amplitude motion[18, 19]. If the large amplitude motion may be described by one parameter, q, the probability distribution of an individual distance carrying out such a motion, may be expressed as[18, 20–22]

$$P(r) = \int P_{fr}(r,q)\, P(q) dq \qquad (13)$$

$P_{fr}(r,q)$ is related to the framework vibration and represents the probability distribution of the individual distance of a hypothetical molecule with a fixed value for q. P(q) is the large amplitude probability distribution, which in a classical approximation may be expressed as

$$P(q) = N \exp(-V(q)/RT) \qquad (14)$$

The actual treatment of the large amplitude motion will thus depend upon the nature of V(q), which is to be discussed in the next section.

3 Large Amplitude Potential Functions

In most cases no exact mathematical expression is known for the potential function, V(q), which is therefore usually approximated by a conveniently chosen series expansion.

In some cases, as for example for large amplitude bending motion, a power series is often used,

$$V(q) = \sum_k a_k q^k \qquad (15)$$

where the a_k-s are the potential coefficients. Negative values for k sometimes are included. For periodic potentials, as for example for torsional motion, a Fourier expansion may be convenient,

$$V(\phi) = \frac{1}{2} \sum_k [V_k (1 + \cos k\phi) + V'_k (1 + \sin k\phi)] \qquad (16)$$

where ϕ is the torsional angle. V_k and V_k' are constants and the k-s are positive integers.

In practice, the potential is approximated only by one or a few terms. This is due to lack of pertinent experimental information, and the quality of the thus obtained potential functions depends upon the convergency of the series. In addition to the two typical potential functions of Eqs. (15) and (16), other mathematical expressions have been used, depending on the problem at hand and the amount and quality of the available experimental data to which the potential must comply.

The choice of series is not only dependent on the type of molecular motion. For example the power series may be convenient for an accurate description of the potential function close to the minima, while a Fourier series is convenient for describing potential barriers to torsional motion.

The electron-diffraction method is not well suited for suggesting what kind of series or what kind of mathematical function should be chosen for a given molecular problem[22, 23]. In the case of C_3O_2[12c, 24, 25], for example, the electron-diffraction method is able to exclude the possibility of a pure quadratic term, but is not able to distinguish between the two potential functions which have been suggested:

$$V(\alpha) = V_2\, \alpha^2 + V_4\, \alpha^4 \qquad (17)$$

and

$$V(\alpha) = A\alpha^2 + \frac{B}{C+\alpha^2} \qquad (18)$$

where α is half the bending angle and V_2, V_4, A, B, and C are constants. Although apparently quite different, $V(\alpha)$ from Eqs. (17) and (18) may be parameterized to have similar shapes for moderate values of α. The great difference between the two functions then obtained at large α-values cannot be recognized by electron-diffraction, since the probability of finding large α-values is vanishingly small.

In open chain molecules the torsional angles are the obvious independent variables for the potential function. But even in the simplest case, that is ethane, the origin of the torsional barrier is not fully understood, though *ab initio* calculations represent the experimental barrier fairly well[26]. For larger molecules theoretical predictions for the potential functions are often based upon semiempirical molecular mechanics calculations[27, 28].

The description of the potentials in open chain molecules is almost always approximated by Eq. (16), often simplified due to molecular symmetry. $V(\phi)$ is in many interesting cases symmetrical about $\phi = \pi$, in which cases Eq. (16) gets the form

$$V(\phi) = \frac{1}{2} \sum_k V_k (1 + \cos k\phi) \qquad (19)$$

For ethane itself and for other molecules with three-fold symmetry k-values other than 3, 6, 9, etc., will vanish.

Open chain molecules have been widely studied using the electron-diffraction method and with considerable success. But quantitive barrier calculations meet with substantial difficulties. In cases with torsional barriers higher than, say 4 kJ/mol, the electron-diffraction method provides information mainly on the regions of the potential function near the minima. For lower barriers the method is usually not sufficiently sensitive to changes in the assumptions on $V(\phi)$. If the barrier is, say 2 kJ/mol or less, the electron-diffraction results may in many cases be indistinguishably like free rotation. Accordingly in order to use the electron-diffraction method successfully for the study of torsional motion, support as to the choice of potential functions may favorably be obtained from other methods, as for example from microwave spectroscopy.

Ring puckering motion in cyclic compounds involves both valency-angle bending and torsion about single bonds. In a conformer of a cyclic compound the equilibrium may be considered as a result of compromise between bond-angle strain, repulsion (perhaps also attraction) between non-bonded atoms or groups, and degree of bond eclipsing. Usually non-planar conformations are preferred, and often the potential barrier is at least 4 kJ/mol. This implies that a "rigid" model approach, assuming small harmonic puckering amplitudes, sufficiently accurately reproduces the electron diffraction data. In four-membered, saturated rings the puckering may be treated as a one parameter bending motion, usually assuming the quadratic and quatric terms of the power series to be the dominating ones[29]. The degree of non-planarity is then described by a ring-puckering coordinate being zero in the planar form.

For larger cyclic compounds the ring-puckering has to be described by at least two parameters. In five-membered rings the barriers separating the minima in the potential may be so small that a so-called pseudo-rotation takes place. The ring-puckering has been described by a perpendicular displacement coordinate depending on two parameters[30, 31].

For six-membered rings the barriers separating the minima in the potential are as a rule so high[28] that the electron-diffraction data may be satisfactorily interpreted by the assumption of mixtures of "rigid" conformers.

For rings larger than the six-membered rings the problem is more complex, since it may be difficult to distinguish between large amplitude motion like the situation in many five-membered rings and a mixture of two or more conformers as in cyclohexane derivatives.

For optimal use of the electron-diffraction method in large amplitude-motion studies, it is important to take advantage of the knowledge concerning potential functions as obtained by spectroscopical methods. Some features of the spectroscopically obtained findings are given in the next section.

4 Some Selected Barrier Determinations

Spectroscopic methods are now so refined that rather accurate determinations of barrier heights and shapes have been made in many cases. Since electron diffraction generally cannot compete with the best quantitative spectroscopic barrier determinations, it was felt appropriate to assess some selected spectroscopic works in order to learn more about the nature of barriers.

Spectroscopists also generally fit their data to potential functions of the general form of Eqs. (15) and (16) or some other convenient mathematical expression, and the potential coefficients (the a_k–s or V_k–s) are thereby determined. Not all of these coefficients need to be of importance. Some may even vanish for reasons of symmetry. Naturally, it would be desirable to be able to determine those coefficients on which the potential function primarily depends. Particularly for the methyl groups, accurate barriers are now available. The situation is not this fortunate in other cases. Often, rather limited experimental data are accessible. The quality of some of this material may be poor or even ambiguous, and rather arbitrary assumptions are now and then made. As a result, only the first few potential coefficients, often of low precision, are obtained. In fact, even today cases where more than four potential coefficients have been deduced are exceptional.

Equations with only one variable, for example a dihedral angle, are often used. However, it is a well known fact that further structural parameters may and often do change with this selected variable. This phenomenon is generally termed structure relaxation. Little is known about relaxation, but it is believed to be of importance in molecules where steric repulsion, conjugation, change of hybridization, lone pair interactions, intramolecular hydrogen bonding, etc., come into play for certain values of the variable. Relaxation is probably often large in the barrier maxima regions, where repulsion, loss of conjugation, etc., may prevail. The molecular population near barrier peaks is low in most cases, making it difficult to determine experimentally the structure of the relaxed molecule by any available method. Therefore, in cases where structure relaxation predominates, a one-variable approach may lead to erroneous results, and it is presumed that barriers derived this way are particularly dubious.

In barrier determinations of open-chain molecules, Eq. (16) is generally used. Most barriers so far determined have two-fold or higher symmetries and many of the V_k–s of Eq. (16) will consequently vanish for symmetry reasons. Generally, this simplifies the problem and makes it easier to obtain more reliable determinations of the individual potential coefficients. If is also presumed that the role of the elusive relaxation effect would be easier to assess with symmetrical barriers than in the more complicated cases.

4.1 Barriers with Two-Fold Symmetry

For these molecules only V_2, V_4, . . . , etc., of Eq. (16) apply. There are not many molecules with this kind of barrier that have hitherto been investigated, and practically all of them are either ethylene or benzene derivatives. Conjugation along the bond connecting the rotating group with the ethylene or benzene parts is present in their equilibrium conformations. There is normally no such conjugation at the barrier maxima, which for symmetry reasons occur with the rotating groups perpendicular to the ethylene part or, respectively, the benzene ring. Structure relaxation is therefore likely to occur in these molecules and may be suspected to be of considerable importance.

Nitroethylene has been thoroughly investigated[32] by the far-IR technique. $V_2 = -19.650 \pm 0.047$ kJ/mol and $V_4 = 1.05 \pm 0.02$ kJ/mol were determined for a model with no relaxation of the nitro group. By relaxing the nitro group in a manner that closely reproduces microwave observations on vibrationally excited states[33], $V_2 = -20.223 \pm 0.050$ kJ/mol and $V_4 = 1.09 \pm 0.02$ kJ/mol were found. The authors[32] conclude that the potential is well reproduced 50° about its planar equilibrium position, but that the data are still insufficient for an accurate determination of the barrier height.

It is interesting to note that the V_4 term is positive. This means that the potential bottom becomes broader and the peak sharper than what would have been found using only the V_2 coefficient.

In another ethylene derivative, $CH_2{=}CHBF_2$[34], V_2 was determined as -17.5 kJ/mol from IR measurements of the torsional frequency.

A deuterated species of phenol, C_6H_5OD[35], has been investigated by far-IR and microwave spectroscopy and the first two Fourier coefficients determined as $V_2 = -16.19 \pm 1.19$ kJ/mol, $V_4 = 0.24 \pm 0.24$ kJ/mol. In the parent species of phenol[36] a determination assuming only a V_2 term yielded $V_2 = -14.439$ kJ/mol.

V_2 terms of several other compounds have been determined as low as $-(2-4)$ kJ/mol in p-fluorostyrene[37], -20.6 kJ/mol in C_6H_5CHO[38], -18.7 ± 1.9 kJ/mol in C_6H_5CFO[39], -16.3 ± 0.4 kJ/mol in C_6H_5NO[40], -12 ± 6 kJ/mol in $C_6H_5NO_2$[41], and -13.33 kJ/mol in $C_6H_5BF_2$[42]. In none of these have higher order terms or relaxation been assessed quantitatively.

4.2 Barriers with Three-Fold Symmetry

For these molecules only V_3, V_6, . . . , etc., are retained. Methyl group barriers are by far the most extensively studied ones of this type and several hundred determinations of V_3 have now been made. Although groups with three-fold axes of symmetry are abundant, limited data exist for other than methyl groups.

Most methyl group barriers have been determined by the microwave frequency splitting method. Generally, only V_3 is fitted and determined with high precision. This method is reviewed in several places[43]. Up-to-date compilations of barriers are made by Starck[44]. Recently, infrared interferometers and laser Raman spectrometers have been made to operate routinely in the far-IR region and successfully applied to the study of torsional vibrations of gaseous molecules. In fortunate cases overtones, hot bands, etc., may be observed by these two methods and important information regarding V_6 obtained. These two techniques will presumably become increasingly important in the future.

Ethane and its halogenated derivatives have played an important part both in electron diffraction and spectroscopy. In Table 1 we have collected 23 of these compounds having a three-fold barrier. The barriers of the series CH_3CH_2X, X = H, F, Cl, Br, and I are seen to increase with the size of the substituent. The V_6 terms are remarkably small for these compounds, even for ethyl iodide. This coefficient is also found to be small for CH_3CHCl_2, CH_3CCl_3, CH_3CHBr_2, and CH_3CBr_3.

Table 1. Potential coefficients of ethane and several of its halogenated derivatives[a]

Molecule	V_3 (kJ/mol)	V_6	Method	Ref.
CH_3CH_3	12.25 ± 0.10	Negligibly small	IR	45)
CH_3CH_2F	14.05	88 J/mol	IR	46)
CH_3CHF_2	13.31 ± 0.09	b	MW	47)
CH_3CF_3	13.3 ± 0.8	b	Raman	48)
CH_2FCF_3	17.6	b	IR	49)
CHF_2CF_3	14.69 ± 0.40	b	MW	50)
CF_3CF_3	16.3	b	IR	51)
CH_3CH_2Cl	15.42 ± 0.05	16 J/mol	MW and IR	46, 52)
CH_3CHCl_2	17.3	$\|V_6\| < 120$ J/mol	Raman	53)
CH_3CCl_3	22.6	$\|V_6\| < 120$ J/mol	Raman	53)
CH_2ClCCl_3	41.8	b	IR[c]	54)
$CHCl_2CCl_3$	59.4	b	IR[c]	54)
CCl_3CCl_3	73.2	b	IR[c]	54)
CH_3CH_2Br	14.92 ± 0.12	$V_6/V_3 < 0.005$	MW and Raman	55, 56)
CH_3CHBr_2	18.1	$\|V_6\| < 120$ J/mol	Raman	53)
CH_3CBr_3	24.2	$\|V_6\| < 120$ J/mol	Raman	53)
CH_3CH_2I	15.4 ± 0.4	$V_6/V_3 < 0.005$	Raman	56)
CH_3CHFCl	18.0 ± 1.2	b	MW	57)
CH_3CF_2Cl	18.4 ± 0.4	b	MW	58)
CH_3CF_2Br[d]	23.4	$\|V_6\| < 120$ J/mol	Raman	53)
CH_3CClBr_2	23.8	$\|V_6\| < 120$ J/mol	Raman	53)
CF_3CF_2Cl	21.3	b	IR	59)
CF_3CF_2Br	23.1	b	IR	59)

[a] Gas phase values unless specified.
[b] Contribution from V_6 term neglected in barrier calculation.
[c] Solution.
[d] Upper barrier height.

Calculations using plausible bond lengths and angles indicate that at the eclipsed position, which is also the barrier maximum, non-bonded distances which are sterically unfavorable probably exist within the chlorine, bromine, and iodine derivatives. Assuming no structure relaxation in the eclipsed form, the non-bonded H ... Cl and H ... Br distances are found to be roughly 50 pm shorter than the sums of the van der Waals radii of hydrogen and chlorine or, respectively, bromine atoms[60)]. The H ... I distance is correspondingly about 70 pm shorter than the sum of the van der Waals radii of hydrogen and iodine atoms. In the derivations of the small V_6 terms of the chlorine[46)], bromine[56)] and iodine[56)] derivatives, neither V_9 nor relaxation was taken into account.

Similar computations indicate that non-bonded H ... H, H ... F, or F ... F distances for eclipsed forms of the ethane series are not seriously in conflict with steric requirements. Typically, these distances were less than 20 pm shorter than the sum of the van der Waals radii of the atoms in question.

The effect of steric repulsion is not only seen in the mono-halogen series of ethane, but is evident from the hexahalogen derivatives as well. In CF_3CF_3 the barrier height is 16.3 kJ/mol[51], somewhat higher than in ethane, 12.25 ±0.10 kJ/mol[45]. There is probably small steric repulsion in either molecule in their eclipsed forms. In solution a V_3 of 73.2 kJ/mol has been determined[54] for CCl_3CCl_3, about six times higher than in ethane. In this molecule, it is very likely that large non-bonded repulsions between the chlorine atoms will exist in the eclipsed form. The V_3 term of 73.2 kJ/mol of hexachloroethane should not be confused with the real barrier height, as this value has been derived employing only the V_3 term for barrier determination. In this molecule where steric repulsion apparently plays an important role, the application of only one term and neglecting relaxation is probably a gross oversimplification, although it is often the only thing that can be done. The barrier of hexachloroethane is, however, undoubtedly large, and may, fortuitously, be about 73.2 kJ/mol. Only future work can decide the shape and size of this barrier. A similar situation presumambly exists in CH_2ClCCl_3 and $CHCl_2CCl_3$.

Determination of V_3 and V_6 Fourier coefficients have been made for several other molecules possessing methyl groups. Selected examples are displayed in Table 2. In all but the notable case of m-fluorotoluene[65], the V_6 term comes out comparatively small, usually less than 5% of V_3. Mostly, it is found to be positive. Determination of structure relaxation has been attempted in CH_3OH[66], CH_3CHO[62], $CH_3CH{=}CH_2$[46], and $CH_3CF{=}CH_2$[46], but invariably found to be small, if not neglible, and difficult to determine unambiguously[62].

m-Fluorotoluene[65] represents a special case. In toluene[67], which for symmetry reasons has a six-fold barrier, V_6 was found to be as small as −58.37 ± 0.08 J/mol. In o-fluorotoluene V_3 was determined as 2717 J/mol[68], while two sets of data fit the microwave spectrum of m-fluorotoluene as shown in Table 2. This is presumably the only known case for methyl barriers where the V_3 and V_6 terms are of similar magnitude.

Table 2. V_3 and V_6 terms of rational barriers of methyl groups of some selected molecules

Molecule	V_3 (J/mol)	V_6 (J/mol)	Ref.
CH_3OH	4439 ± 36	8[a]	
	4464.3 ± 0.8	6.3 ± 1.7[a]	61)
CH_3CHO	4786 ± 25	130 ± 4	62)
$CH_3CH{=}CH_2$	8289	185	46)
$CH_3CF{=}CH_2$	9627	−13	46)
CH_3NH_2	8201	31	63)
CH_3SiH_3	6837 ± 71	331 ± 75	64)
m-Fluorotoluene	190 ± 13	−95.4 ± 3[b]	
	203 ± 13	63.2 ± 2.1[b]	65)

[a] Depending on approximation made.
[b] Both sets of potential coefficients fit experimental data.

There are few examples where barriers of the CX_3-type where X = F, Cl, Br, and I, have been accurately determined. Besides those listed in Table 1, V_3 of CF_3CHO has been determined as 3.807 kJ/mol[69], somewhat less than 4.786 ± 0.025 kJ/mol found for CH_3CHO[62]. The barrier of CF_3NO of 3.222 kJ/mol[70] is also slightly less than 4.7568 ± 0.0046 kJ/mol determined for CH_3NO[71].

Very few, if any, accurate determinations of V_9-terms have been made. This coefficient is very important in the vicinity of the three-fold barrier maximum, but nearly negligible near the bottom of the potential well. Since most barrier determinations have been made near the potential minimum and extrapolated to the barrier top, it is not unreasonable that the omission of V_9 may be serious at least in some cases.

4.3 Barriers with Six-Fold Symmetry

There are comparatively few molecules possessing four-fold and five-fold symmetrical rotational barriers. Six-fold barriers are more common and for them only V_6, V_{12}, . . . , etc, apply.

Most V_6-barriers determined involve methyl groups, and V_6 is invariably found to be small in corroborations with the findings above. Typical values are: V_6 = −24.400 ± 0.024 J/mol for CH_3NO_2[72], V_6 = −57.61 J/mol for CH_3BF_2[73], and V_6 = −58.37 ± 0.08 J/mol for $CH_3C_6H_5$[67]. For other p-toluene derivatives V_6 is typically about −60 J/mol[44].

N-methyl pyrrole[77] represents an exception. Here, a potential function using V_6 = −549.3 J/mol and V_{12} = −200 J/mol fits the microwave data just as well as a set employing V_6 = −558.6 J/mol and V_{12} = 167 J/mol. The lesson to be learnt from this is that even in cases where the Fourier expansion starts with a high a term as V_6, rapid convergence is not always ensured.

V_6 terms have been determined for some compounds not possessing methyl groups. In CF_3NO_2 V_6 = −311 J/mol[75]. Although this is a small barrier, it is more than ten times the value found for CH_3NO_2[72]. In $CF_3C_6H_5$ a value of −42.7 J/mol was found[76], which is fairly similar to the toluene value[67]. In $SiH_3C_6H_5$ V_6 = −74.39 ± 0.08 J/mol[77] which is also near the toluene barrier height[67]. A very small barrier of V_6 = −7.9 ± 3.3 J/mol was found for SiF_3BF_2[78].

At least one twelve-fold barrier has been determined to date. In the "cage"-molecule $CH_3B_5H_8$ V_{12} was found to be less than −12 J/mol[79].

4.4 Barriers with Low Symmetry

Barriers with low symmetry are usually much more complicated than other types. Several Fourier coefficients are normally necessary. Sufficient high-quality experimental data to derive the necessary Fourier terms are often very difficult to obtain. Despite these severe obstacles, several such barriers have been determined in recent years[44]. Some few of these potentials are presumably of high quality and represent, indeed, great experimental and intellectual achievements.

Due to the complexity of the problem of low symmetry barriers, influence from relaxation and convergence properties of the Fourier expansion are difficult to extract from the present material. However, in the case of 3-fluoropropene[80] which has one of the best known rotational barriers of this complicated kind, the Fourier series do not converge rapidly. It was possible to determine the first six potential coefficients as $V_1 = -2.95 \pm 0.36$ kJ/mol, $V_2 = -2.21 \pm 0.30$ kJ/mol, $V_3 = 10.25 \pm 0.17$ kJ/mol, $V_4 = -2.25 \pm 0.30$ kJ/mol, $V_5 = 0.08 \pm 0.06$ kJ/mol, and $V_6 = -1.11 \pm 0.12$ kJ/mol. V_6 is thus seen to be of the same order of magnitude as V_1, V_2, and V_4. It remains to be seen whether slow convergence should generally be expected.

4.5 Conclusion

The overwhelming part of accurate barrier determination performed to date represents methyl group barriers. Except for notable cases like m-fluorotoluene[65] and N-methyl pyrrole[74] the first V_k of Eq. (16) alone seems to give a remarkably good representation of both the potential shape and barrier height for this group. Relaxation appears to be a rather small effect. The same seems to be true for the CF_3 group, and SiH_3 and GeH_3 groups can presumably be expected to behave likewise. Barriers of methyl and CF_3 groups are usually fairly similar in identical environments.

The simple situation encountered for methyl groups should not lead us into believing that this is generally the case. Relaxation may be of great importance as was seen for nitroethylene[32], and there is no *a priori* reason why expressions such as Eq. (16) should indeed converge rapidly. Barrier determination of more complex groups than methyl will presumably remain a challenge in the years to come. This task should be approached with a maximum of caution and a minimum of prejudice.

5 The Shrinkage Effect

The term shrinkage was introduced to describe and explain a phenomenon first observed for the linear molecule dimethyldiacetylene[17a], allene[17b], carbon suboxide[81, 82], butatriene[83], and carbon disulphide[84]. For these molecules the long internuclear distances, as studied by electron diffraction, are observed shorter than corresponding to the value obtained by adding the observed individual bond distances. If a static molecular model is used, the molecule seems bent. The effect was qualitatively explained as a result of molecular vibration, particularly out-of-linearity vibration. The shrinkage effect is of course not restricted to linear molecules. The shrinkage of a distance is defined as the difference between the distance as calculated from a rigid geometric model with the observed bond distances and valence angles, and the distance as observed directly by electron diffraction. Usually the shrinkage effect is small and does not reveal large amplitude motion. However, the effect is large enough to cause significant errors in structure parameter determination by electron diffraction if not properly allowed for. For example the shrinkage of the longest C . . . C distance in allene[17b] and butatriene[83] was observed to be 0.6 pm and 1.3 pm respectively. For carbon suboxide, however, the shrinkage was observed to be of an order of magnitude larger[81, 82]. For the O . . . O distance the shrinkage was observed to be 15 pm. This molecule has, therefore, attracted great interest by electron diffractionists, spectroscopists, and theoreticians.

The shrinkage effect is now quantitatively well understood for small amplitude vibrations[17c), 85]. The phenomenon is thoroughly studied by Cyvin in a series of articles. The result of his own work and that of others is described in detail in Cyvin's book[4].

Large amplitude shrinkage is of course the main interest in the present context, but again the question about the dividing line between small and large amplitude motion arises. Even in the simplest kind of molecules where shrinkage is to be expected, namely the linear triatomic molecules, cases are known with very large angular bending motions, of course particularly at high temperatures. For molecules like $MnCl_2$, $FeCl_2$, $CoCl_2$, $NiCl_2$, and $NiBr_2$ measurements at about 800 °C lead to shrinkage values of approximately 10 pm which corresponds to an angular shrinkage of about 25°[86]. For CaI_2 and SrI_2 the shrinkage effect was observed as large as 22 pm and 29 pm, respectively[87]. The temperatures were 1300 °C for CaI_2 and 1250 °C for SrI_2. This corresponds to a shrinkage in angle of 32 ° and 36 °, respectively. In such cases it is obvious that the question of the relation between the temperature dependent "average structure" of a molecule, as determined by electron diffraction, and the "equilibrium" or "zero-point average" structure is difficult to deal with. This problem has in particular been studied by Kuchitsu[88] based upon the fundamental work of Bartell[89].

For triatomic molecules with large shrinkage it may be difficult to decide whether the molecule is actually a linear molecule or bent. Semantically the difference between a linear and bent molecule ought to be clear and defined by the minimum in the potential function of the molecule. But on the other hand electron-diffraction studies cannot easily distinguish between a symmetrical double minima potential near 180 ° and a potential with one single minimum at 180 °, corresponding to a genuine linear molecule. The question of linearity has in many cases to be left to other methods as for example measurement of electric polarity in high temperature vapor by electrical deflection of molecular beams with mass spectrometric detection[90–92].

Though the shrinkage effect obscures the structure determination of a molecule, it may contribute with information as to the flexibility of the molecule. For a molecule of the type MX_2 with linear equilibrium configuration it is possible to estimate the bending vibrational frequency, ν_2, from measured shrinkage values[86]. For a triatomic molecule with a shrinkage of approximately 10 pm the ν_2-value is estimated to be about 1 kJ/mol, in good agreement with matrix-isolation infrared spectroscopic studies[93].

As earlier indicated carbon suboxide has attracted considerable interest as to the study of shrinkage effect and large molecular motion. The molecule has been submitted to four independent electron-diffraction studies at this laboratory[81, 82, 12c]. The large shrinkage effect of carbon suboxide could not be explained from the known vibrational frequency at the time when the shrinkage effect was first observed. An unobserved bending frequency (ν_7) was predicted and was later observed to be about 750 J/mol[94, 95].

The last electron-diffraction study of carbon suboxide included measurements at three temperatures (237 K, 290 K, and 508 K). The measured shrinkage data matched the theoretically calculated values based on the measured ν_7 value[96], both as to the value for the shrinkage itself and as to the temperature dependence. (It should be remembered that temperature is rather ill-defined in an electron-diffraction gas experiment.) It was demonstrated that the C=C=C-bending vibration, which is the main contributor to the large shrinkage effect, could not be harmonic. Doubts about the actual linearity of the molecule were revived and the possibility of a double minimum potential function for the ν_7 bending with a barrier from about 0.5 kJ/mol to 3 kJ/mol seemed likely. The minima were estimated to be at about 12 ° to 16 °, described by the angle α in Fig. 2. This corresponds to a C=C=C angle of 156 ° to 148 °. Two further studies based upon electron-diffraction and spectroscopy[24, 97] support these assumptions, indicating a barrier of about 0.6 kJ/mol, though a non-barrier potential could not be ruled out[1].

Several rather conflicting *ab initio* calculations have been reported, one supporting a linear equilibrium conformation with a nearly harmonic potential[98], another one producing a barrier of 8 kJ/mol and minima at the α value of 27.5°[99]. Neither of these results is in accordance with electron-diffraction findings. A more recent *ab initio* calculation failed to produce a double minimum potential curve[100], but verified the high degree of anharmonicity in the C=C=C-bending potential. Several spectroscopic studies have been carried out with results in good agreement

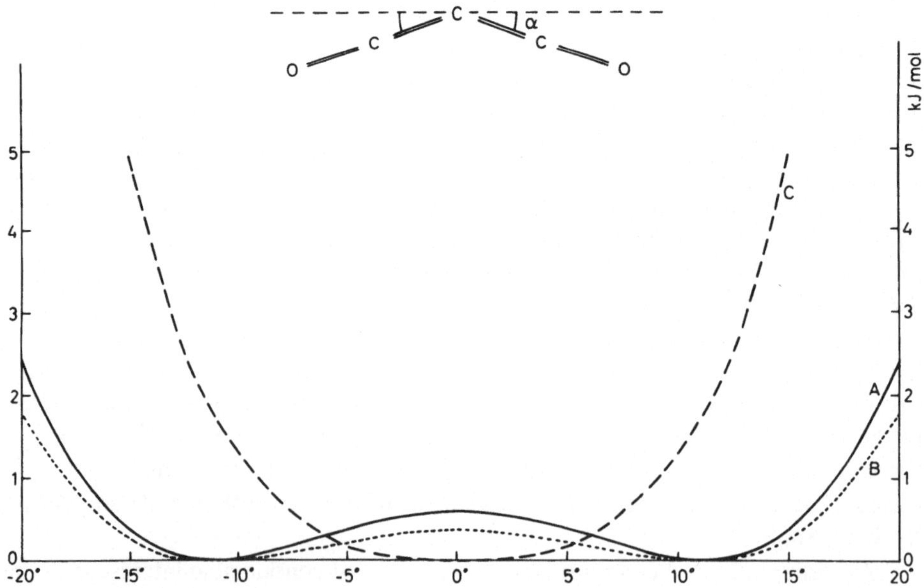

Fig. 2. Potential curves for C_3O_2, estimated from electron diffraction (A), spectroscopical data (B) and *ab initio* calculations (C)

1 See Note Added in Proof.

with electron-diffraction data[101–103], producing α values ranging from 11° to 13° and barrier hights ranging from 170 to 700 kJ/mol.

The situation as it looks at present is summarized in Fig. 2 where the latest electron-diffraction barrier is compared with the latest *ab initio* calculation and one of several spectroscopically based calculations[103]. The two dimentional presentation of the potential curve may of course be considered as a cross section of a three dimensional potential surface with a circular valley.

The contrasting results demonstrated in Fig. 2 calls for further endeavor and combination of various methods. Particularly *ab initio* calculations including larger bases as already suggested[100], may be profitable.

A series of molecules containing the group Si–N=C=O or Si–N=C=S presents problems similar to C_3O_2. Electron-diffraction data for H_3Si–N=C=O and H_3Si–N=C=S, conventionally interpreted in a rigid molecule analysis, leads to a Si–N=C angle of 152° for the former and 164° for the latter of the two molecules[104]. The electron-diffraction study further shows that the SiNC bending mode had an unusual large amplitude even at a temperature of 0 °C. For the Si–N=C=O chain a small potential hump of approximately 240 J/mol at the linear configuration is suggested, while a harmonic potential is suggested for the Si–N=C=S chain. A detailed microwave study of H_3Si–N=C=O[105] confirms this finding producing a value for the α-angle of 11° and a barrier of 272 J/mol. The ground vibrational state is found almost exactly on the maximum of the barrier.

An X-ray study of crystalline H_3Si–N=C=O at 140 K is not consistent with any large amplitude bending motion[106]. Apparently the packing forces in the crystal constrain the low frequency bending vibration of the free molecule. The Si–N=C angle is found to be 158.2 °, corresponding to an α value of 10.9 °. Since the energy required to bend the molecule is so small, it is not surprising that lattice energy may interfere with the intramolecular motion.

6 Internal Rotation in Open Chain Molecules

6.1 Ethane Like Molecules

The factors influencing the conformational stability in open chain molecules have previously been treated extensively in review articles (see for example Ref.[6] and [107]). The aim of the present section is to study torsional potential functions of a series of molecules of principal importance, in particular related to the results of electron-diffraction investigations. The bulk contents of information obtainable from an electron-diffraction intensity curve of a molecule carrying out torsional motion, are not concerned with the torsional motion at all. The part of the intensity curve giving information about the torsion, is distributed over the same range of the intensity curve as where the torsional independent information may be obtained. In the RD-curve the contribution from the torsional dependent part is more clearly separated. To illustrate this and the general influence of torsional motion, three simple molecules with three-fold torsional barriers have been selected (Figs. 3–5)

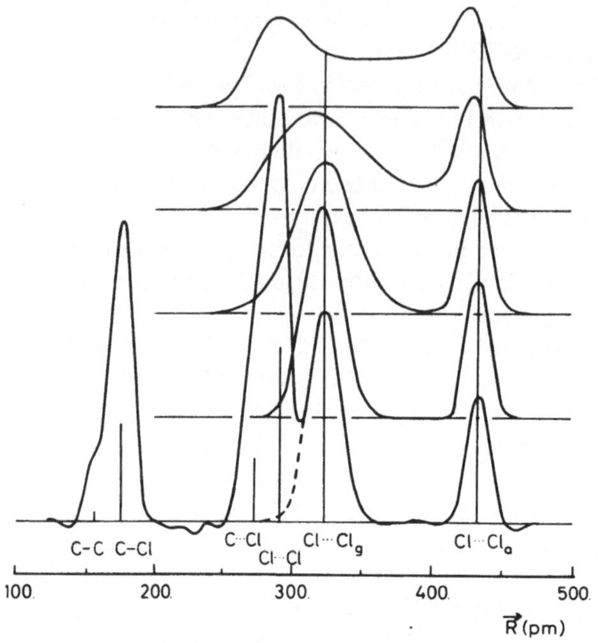

Fig. 3. Radial distribution curves for hexachloroethane. The vertical lines give the Cl · · Cl positions in *gauche* (*g*) and *anti* (*a*). Curve A is experimental, the dashed line combined with the other part, indicates the torsional dependent contribution, obtained by subtracting the theoretical torsional insensitive part from the experimental curve. Curves B–E are theoretical torsional dependent distribution curves. (B) based on a rigid, staggered model with u_g = 14.3, u_a = 6.7 (pm). (C–E) calculated for large amplitude models, using framework vibrations and a torsional potential $\frac{1}{2}V_3(1 + \cos 3\phi)$ with V_3 equal to 12.5, 4.2, and 0 (kJ/mol), respectively. The scaling between A and the other curves is somewhat arbitrary, and the damping factors and modification functions slightly different

The molecules chosen are hexachloroethane[108, 109], hexachlorodisilane[109], and hexafluorobutyne-2[110]. Curve A in each of the three figures is the experimental RD-curve. The outer part of the curve, in combination with the dashed line, defines the *only* torsional dependent contribution, namely the long halogen-halogen distance. Curves B–E are theoretical curves for the same torsional sensitive distances, calculated under different assumptions. Curve B is calculated for a rigid, staggered model using spectroscopically obtained u-values and harmonic torsional vibration. Curves C–E are calculated differently using Eqs. (13) and (14), assuming barrier heights 12.5, 4.2, and 0 kJ/mol, respectively. The individual curve is thus composed of a weighted sum over molecular species for all values of ϕ, each species undergoing small, harmonic framework vibrations, spectroscopically estimated[19, 111]. The different species are weighted according to the normalized classical Boltzmann coefficients which depend directly on the hindering potential $V(\phi)$. For symmetry reasons $V(\phi) = \frac{1}{2}V_3(1 + \cos 3\phi)$, higher terms ($V_6$, V_9 etc.) being neglected. For each of the three figures *gauche* and *anti* positions are indicated.

Comparing the curves visually, it is seen that the hexachloroethane molecule is satisfactorily described by a rigid, staggered model. Hexafluorobutyne-2 may be

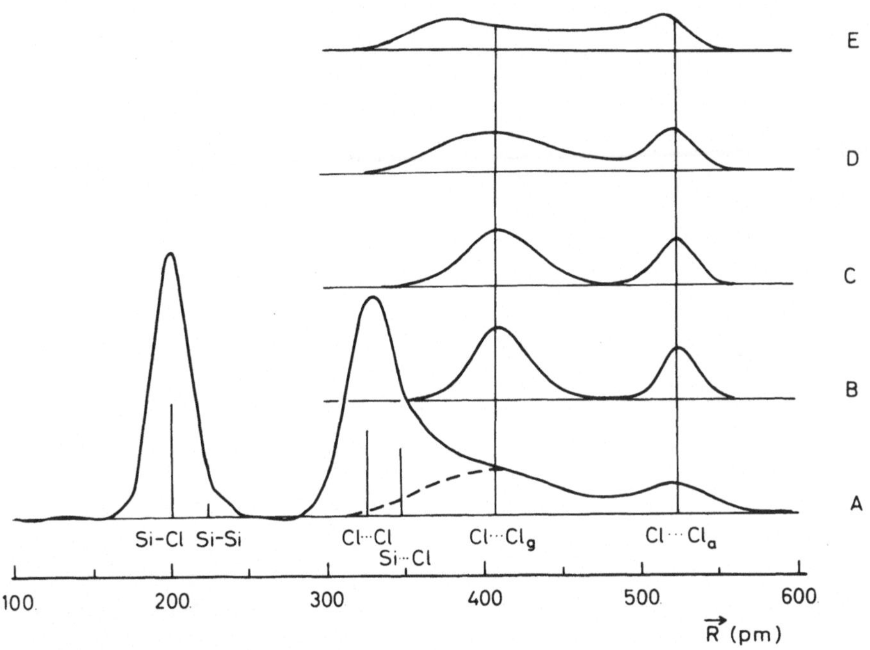

Fig. 4. Radial distribution curves for hexachlorodisilane. See caption for Fig. 3. $u_g = 19.0$, $u_a = 9.9$ (pm)

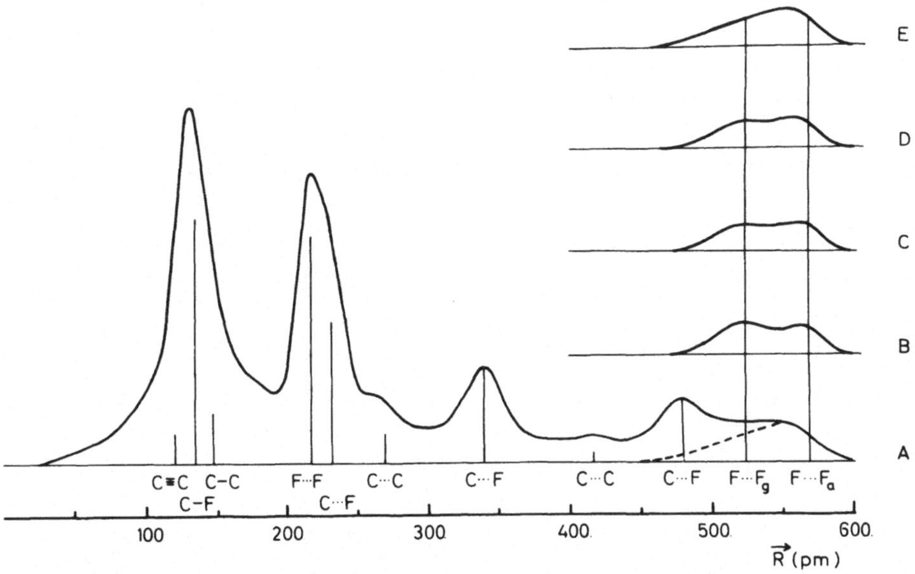

Fig. 5. Radial distribution curves for hexafluorobutyne-2. See caption for Fig. 3. $u_g = 21.1$, $u_a = 11.6$ (pm)

Table 3. Some torsional dependent observed quantities for molecules with symmetric end-groups. Central distance (R) and u-values in pm, torsional amplitude (σ_ϕ) in °, Fourier term (V_n) in $kJ mol^{-1}$

	ED								Other methods	
Molecule	R	σ_ϕ	u_g	u_g^{fr}	u_a	V_n	Remarks[a]	Ref.	V_n	Ref.
CH_3–CH_3	153.2						Assumed staggered	112–114)	12.2 (IR)	45)
CF_3–CF_3					6.9	18.0	u_g, $u_a \approx u_g^{fr}$	20)	16.3 (IR)	51)
	154.5	7.3			6.7	28.4	δ_a	115)		
			12.2	9.0		16.3	u_g, u_g^{fr}	115)		
			13.0[b]	9.4[b]	6.3[b]	15.1	Weighted sum	115, 19)		
CF_3–CF_2I	152.3						Assumed staggered	116)		
CCl_3–CCl_3	156.2		13.6		10.9	45.1	u_g, u_g^{fr}	18)	73.2 (IR(sol))	54)
					8.0	53.5	u_g, $u_a \approx u_g^{fr}$	20)	62.7 (IR(g))	117)
			10.1		6.5	61.5	u_g, $u_a \approx u_g^{fr}$	20, 108)		
			14.3[b]	11.6[b]	6.7[a]			19)		
CBr_3–CBr_3	154.		19.0		11.0			118)		
			13.6[b]	11.9[b]	6.6[a]			19)		
$C(CH_3)_3$–$C(CH_3)_3$	158.2	5.0	13.8		7.7		δ_a	119)		
BCl_2–BCl_2	170.2	20–30	22.3	8.0		7.7	Weighted sum	12d)	7.1 (IR(g))	120)
						6.1	Weighted, $V(\phi) = \frac{1}{2} f\phi^2$	12d)		
CCl_3–SO_2Cl	186.5	9.6	19.0	8.8	8.2	16.3–23.4	u_g, $u_a \approx u_g^{fr}$	121)		
						18.0–22.3	u_g, u_g^{fr}	121)		
						14.6–17.6	δ_a	121)		
CH_3–SiH_3	186.7								6.8(V_6 = 0.3)(MW)	64)
CCl_3–$SiCl_3$	193.2		19.0		10.0	16.7	Weighted sum	18)		
						18.0	u_g, u_g^{fr}	18)		
						16.3	u_g, $u_a \approx u_g^{fr}$	20)		
CCl_3–$GeCl_3$	198.					13.0	u_g, $u_a \approx u_g^{fr}$	122)		
SiH_3–SiH_3	233.1						Assumed staggered	123)	5.1 (electrostatic)	124)
SiF_3–SiF_3	231.7		15.1		9.4	2.2	Weighted sum	125)		
		25.4	17.3/20.7		14.4		Assumed staggered	126)		

$SiCl_3-SiCl_3$	226.4			17.	11.	4.2	Weighted sum	18)		
						4.6	u_g, $u_a \approx u_g^{fr}$	20)		
	232.6		29.5		14.4		Assumed staggered	127)		
$Si(CH_3)_3-Si(CH_3)_3$	234.0	10.			20.		δ_a	128)		
$C_5H_5-Fe-C_5H_5$	412.8	0.	9.7/15.0		11.9	−4.6	Assumed eclipsed	129)	7.5 (NMR(s))	130)
							u_g, three $\vec{u}$-s			
			11.7/19.3		10.1	−3.8	Weighted sum	22)		
$CH_3-C{\equiv}C-CH_3$	414.9					0.	Weighted sum	131)		
$CF_3-C{\equiv}C-CF_3$	416.0		21.2[b]	20.7[b]	11.5[b]	0.4	Weighted sum	110, 132, 111)		

[a] The given remarks imply the following:
Assumed staggered: a rigid staggered model with reasonable u-values has been applied, but no barrier estimate has been given.
u_g, u_g^{fr}: The barrier has been estimated from the torsional contribution to u_g, obtained by correcting u_g for framework vibration (u_g^{fr}) calculated from spectroscopic data.
u_g, $u_a \approx u_g^{fr}$: The barrier has been estimated from the torsional contribution to u_g and it has been assumed that u^{fr} is independent of ϕ, which leads to the approximation $u_a \approx u_g^{fr}$.
δ_a: The barrier has been obtained from the estimated shrinkage effect in *anti*, δ_a, often expressed as an apparent torsional angle, approximately equal to σ_ϕ.
Weighted sum: a large amplitude model has been introduced according to Eqs. (13) and (14).

[b] Calculated from spectroscopic data.

approximated by a freely rotating model.Hexachlorodisilane represents something in between the two extremes, the barrier, i. e. V_3 when described by the single term cosine potential, is close to 4 kJ/mol and definitely smaller than 12 kJ/mol. The reported estimate is 4.5 kJ/mol[20,18].

It should be emphasized that the theoretically calculated curves (B–E) all are somewhat dependent on spectroscopic data. Furthermore the results obtained using the staggered model approach, may be indistinguishably like the results obtained using a cosine-term potential, if the barrier is high enough. For high barrier cases the contribution to the electron-diffraction data is predominated by molecular species corresponding to the minimum regions of the potential curve. The information thus obtained from electron diffraction to the potential curve near maxima is therefore practically nil in high barrier cases.

The primary aim of the staggered model approach is to describe the curvature of the potential function at the minima. Therefore, in order to use this approach for barrier determination, one has to choose a potential function with a curvature at the minima equal to that derived for the staggered model. The most obvious choice of potential function is again the one-term cosine function. It is therefore no surprise that the two approaches, the staggered model approach and the approach using Eqs. (13) and (14), lead to the same results if they use the same potential function.

Also in the staggered model approach the u-values for the halogen-halogen distances are composed of contribution both from framework vibration and torsional motion. The torsional motion part may be expressed by σ_ϕ, the root-mean-square deviation from the minimum position. For the molecules so far described, the value of σ_ϕ is to a good approximation equal for the *gauche* and *trans* peaks. (This is of course not the case for molecules like 1,2-dihaloethanes).

Since the *anti*-distance varies so little with ϕ, the u-value of the *anti*-distance is primarily due to framework vibration, particularly in high barrier cases[109]. The *anti*-peak is accordingly not suited for direct determination of σ_ϕ. On the other hand, the torsional motion leads to an asymmetry in the *anti*-peak due to the functional relation between r and ϕ. For a low barrier case this asymmetry may be appreciable, while in a high barrier case it may be observed only as a shrinkage effect for the *anti*-distance. The asymmetry or the shrinkage may be used to derive a value for σ_ϕ.

Whether a staggered model will reproduce the torsion sensitive distance distribution with sufficient accuracy for intermediate barriers, or a weighted sum over Gaussian peaks has to be applied, will also depend on the total change in the torsion dependent distance compared to the u-values of the said distance.

In Table 3 are listed some experimental values related to the torsional motion for a series of symmetric rotor molecules. The given examples are chosen to illustrate the points discussed in this section.

Inspection of the table reveals the general consistency between results obtained from either of the two above mentioned approaches for intermediate barriers. The staggered model approach leads to *gauche* vibrational amplitudes (u_g) in satisfactory agreement with values obtained from other methods. The determination of barriers from u_g-values requires that the appropriate u-framework values (u_g^{fr}) are known. These parameters may be obtained from spectroscopy, but if not available,

u_g^{fr} may also be estimated from electron-diffraction data using a rough approximation. In order to do so, one has to assume that the framework vibration is constant during the internal rotation. Since the torsional contribution to the u-value in the *anti* position (u_a) is negligible, the u_g^{fr} may then be approximated by u_a[20, 109].

The low barriers must be determined from Eqs. (13) and (14) in order to obtain acceptable shape of the distance distribution. Since this procedure requires that u-framework is known in the whole ϕ-interval, accurate barrier estimates depend on the quality of the applied u-framework values[110].

Generally, the results given in Table 3 are reasonable judging from the effects to be expected from steric hindrance, both taking the size of the substituents in the rotating groups and the bond length separating the groups into consideration.

A similar analysis of data obtained from molecules with asymmetric end groups is more complicated. Apart from the problems connected with the separability of the torsional motion from the framework vibration, experience shows that several more terms have to be included in the Fourier series to describe the torsional potentials properly. On the other hand, the electron-diffraction data from asymmetric molecules usually contain more information about the potential function than data from the higher symmetric cases. In conformity with the results obtained for symmetric ethanes the asymmetric substituted ethanes, as a rule, exist as mixtures of two or more conformers in the gas phase. Some physical data for asymmetric molecules are given in Table 4. The electron-diffraction conformational analysis gives rather accurate information about the positions of the minima in the potential curve. Moreover, the relative abundance of the coexisting conformers may also be derived. If the ratio between the concentrations of two conformers is equal to K, one may write

$$\Delta G = -RT\ln K = \Delta H - T\Delta S \tag{20}$$

where ΔG, ΔH, and ΔS are the free energy, enthalpy, and entropy differences respectively for the two conformers. Thus, by studying the conformational mixtures at different temperatures, ΔH and ΔS may be derived. To a sufficiently good approximation ΔH represents the energy difference between the two minima corresponding to the conformers involved. Information about the curvature in the minima of the potential curve may also be obtained in a way analogous to that of the symmetric rotor molecules.

A series of molecules of great interest in principle and also well suited for electron diffraction studies are the 1,2-disubstituted ethanes. *Gauche*- and *anti*-conformers coexist in the gas phase in all known cases, sometimes with the *anti*-conformer as the prevailing conformer as in 1,2-dichloroethane[136], sometimes with *gauche* prevalence as in 2-chloroethanol[7, 8].

The 2-chloroethanol may serve as an example of the effect of temperature on conformational equilibrium[7, 8]. The molecule has been studied at five different temperatures. In Fig. 6 the lower curve corresponds to the lowest temperature studied (T = 310 K) and the upper curve to the highest temperature (T = 523 K). The main difference between the two curves is the barely observable *anti* O . . . Cl-peak in the low temperature curve in contrast to the well developed *anti*-peak in the high temperature curve. The *gauche*-peak is correspondingly reduced in the high temperature curve compared to the low temperature curve. In Fig. 7 the observed

Table 4. Some torsional dependent observed quantities for molecules with asymmetric end-groups. Central distance (R) and u-values in pm, the torsional angle for a *gauche* distance type (ϕ_g) in °. ΔH is the enthalpy difference between two conformers in kJ mol^{-1}, and V_n the related Fourier terms in kJ mol^{-1}

Molecule	ED R	ϕ_g	u_g	u_a	ΔH^a (V_n)	Ref.	Other methods ϕ_g	ΔH (V_n)	Ref.
CH_2F-CH_2F	150.5	69.4	11.0	8.0	−5.8→−2.5	133, 134)	50	−8.3(R(l)) (V_{max}[b] =5.6)	135)
CH_2Cl-CH_2Cl	153.1	75.3	14.3	6.6	4.6	136)	65.5	4.6(IR(g)) (V_1= 15.1, V_2= 7.9, V_3= 23.7)	137)
CH_2Br-CH_2Br	150.6	73.	16.5	7.2	9.2	138)		6.7(R(l)) (V_{max}=17.6)	139, 140)
$CH_2(CH_3)-CH_2(CH_3)$	153.1	65.	12.6	7.2	~4.0	141)		3.2(R(l))	142)
$CH_2(CN)-CH_2(CN)$	154.2	70.	18.0	8.0	~6.3	143)		−1.5(IR(l))	144)
$CH_2(CCH)-CH_2(CCH)$	155.2	78.	14.1	7.1	~4.2	145)		(V_{max}=40)	140)
CHF_2-CHF_2	151.8	78.[c]	10.5	6.0	4.9	146)	66.5	4.8(IR(g)) (V_1= 12.3, V_2= 4.9, V_3= 14.9)	137, 147)
$CHCl_2-CH_2Cl$	158.0	75.	14.9	9.8		148)	62.	13.0(IR(g)) (V_1=−22.2, V_2=− 5.5, V_3= 30.3)[d]	137)
$CH(CH_3)_2-CH(CH_3)_2$	154.6	65.	12.2	7.5	0.	119)		0.4	140)
$CH_2(CH_3)-CH_2Cl$	153.5	59.	11.4	6.8	−1.3[e]	149)		−2.1(IR(g))	150)
$CH_2(CH_2Cl)-CH_2Cl$	152.9	65.8	15.6		−4.6[e]	151)			
$CH_2(CH_2Br)-CH_2Br$	152.7	66.6	12.4	9.8	−3.8[e]	152)			
$CHCl(CH_2Cl)-CH_2Cl$	152.6	55.5[f]	14.0		−7.5[e]	153)			
$CHBr(CH_2Br)-CH_2Br$	153.4	55.7[f]	18.1		−6.3[e]	154)			
$CH(CH_3)_2-CH_2Cl$	153.4	66.[f]	11.	8.6	3.4[e]	155)		0.96(IR(g))	140)
$CHCl(CH_3)-CH_2(CH_3)$	153.4	(60)	9.8	7.5	−1.7[e]	156)		−2.10(IR(l))	157)
$CH_2(CH_2Cl)-CH_2(CH_3)$	153.3	(60)	10.8	7.5	0.	156)			
$CH_2(CH_2Br)-CH_2(CH_3)$	153.2	67.		9.0	1.	158)			
$SiCl(CH_3)_2-SiCl(CH_3)_2$	233.8	73.2	24.9	11.5	−2.5	159)			

$CH_2Cl-C{\equiv}C-CH_2Cl$	410.9	(19.4)[g]	11.3	$\begin{pmatrix} V_1 = 1.25 \\ V_2 = -0.04 \end{pmatrix}$ nearly free rotation	160) 162)	$(V_1 < 0.4)$	161)
$CH_2Br-C{\equiv}C-CH_2Br$	414.4	45.4 (24.)[g]	13.3	$\begin{pmatrix} V_1 = 2.1 \\ V_2 = 2.1 \\ V_3 = -2.1 \end{pmatrix}^{h}$	163, 6)	Nearly free rotation, (IR,R(g)) pot. minimum of C_2 symmetry	164)

a Δ = g-a (*gauche-anti*).
b V_{max}: barrier height between *anti* and *gauche*.
c ϕ_g = ∡ HCCH, opposite ∡ FCCF is unreasonably small (42°).

d ϕ = 180° is for the symmetric form 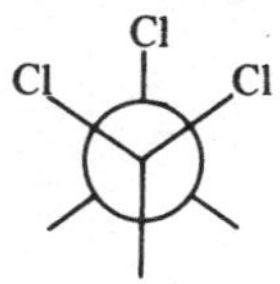

in the Fourier expansion.

e Δ is between *gauche* and *anti* of methyl/halogen positions.
f ϕ_g = ∡ MeCCHal; $\phi_g < 60°$ implies that ∡ HalCCHal > 60°, when two halogenes are present, ϕ_g refers to the *gauche* interaction angle found in the most stable conformer.

g (u_g) is u_g^{fr}.

h Giving a barrier in *anti* equal to 3.1 kJ mol^{-1}, and a pot. minimum at ~100 °.

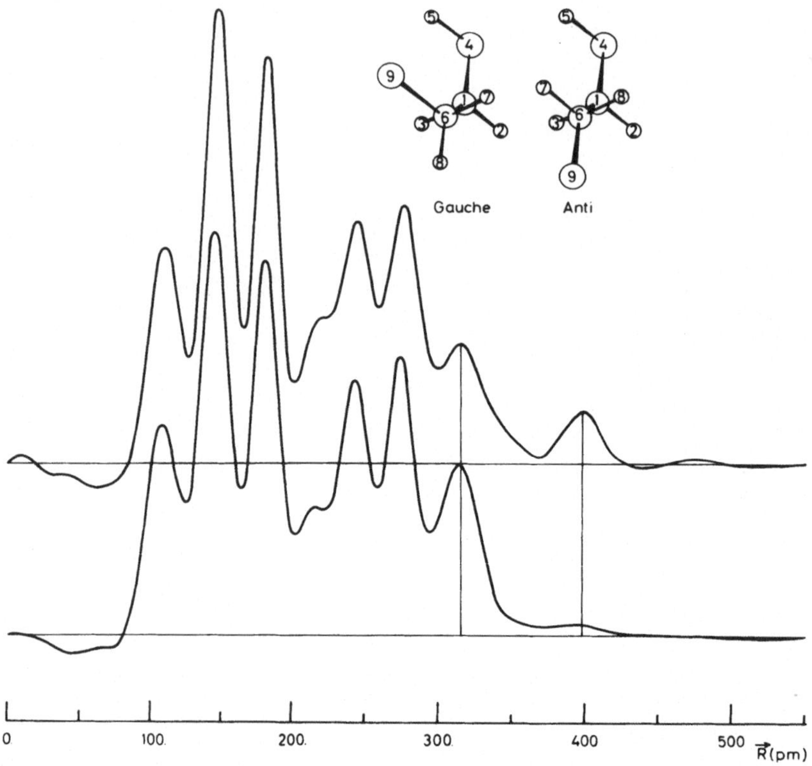

Fig. 6. Experimental radial distrubution curves for 2-chloroethanol. (k = 20 pm^2) at 523 K (upper) and 310 K. The vertical lines indicate the Cl · · · O positions in *gauche* and *anti*

RlnK is plotted against 1/T using Eq. (20). The various points represent results from different ways of deducing K from the electron-diffraction data. From the slope of the best straight line through the observed points ΔH may be obtained, while ΔS is determined by the point of intersection with the ordinate. The values thus obtained for ΔH and ΔS are -10.0 kJmol^{-1} and 11.7 J mol^{-1}K^{-1} respectively.

For a qualitative description of the potential curve corresponding to *anti* prevalence, two terms in the Fourier series [Eq. (19)] seem sufficient, namely the V_1- and V_3-term. In a case with *gauche* prevalence the terms V_2 and V_3 seem sufficient. As an example of these two possibilities may serve 1,2-dichloroethane[136)] and 1,2-difluoroethane[133, 134)]. Plausible potential curves of these two molecules are presented in Fig. 8. Electron-diffraction studies alone are able to demonstrate in both these cases that a two term potential is insufficient to give a quantitative description, i. e. incapable of reproducing the experimental values for ΔH and the torsional angle ϕ_g. In fact a series of V_k-terms is needed. An extensive analysis[165)] based upon electron diffraction indicates that as many as 8–12 terms may have to be included to reproduce the experimental findings with sufficient accuracy. This is to a great extent caused by the rather large deviation of the experimental ϕ_g from the idealized 60 ° *gauche* torsional angle. From Table 4 it may be seen that ϕ_g ranges from about 70 to 75 ° in the 1,2-disubstituted ethanes. In order to produce such a minimum

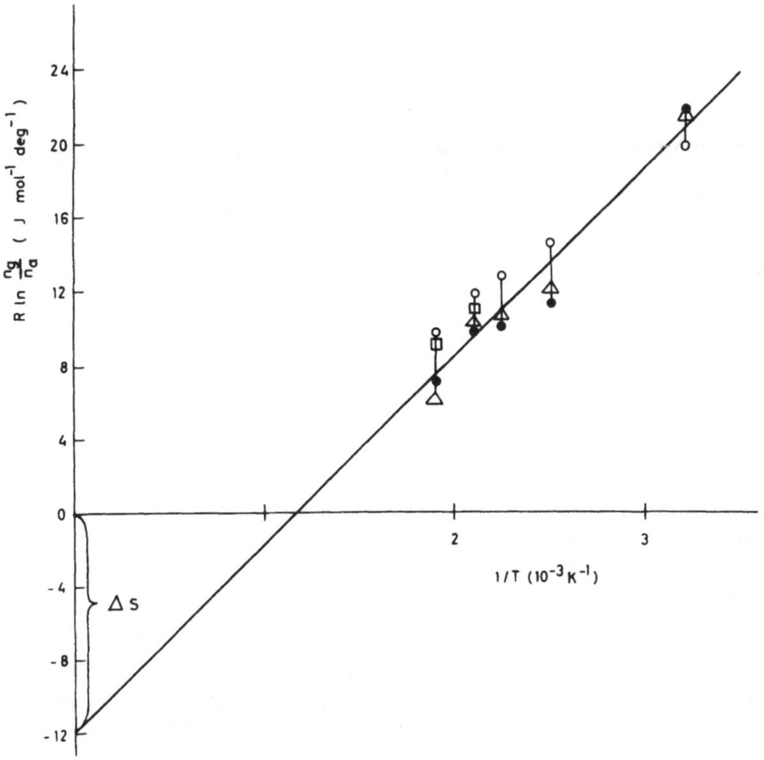

Fig. 7. R ln K for 2-chloroethanol as a function of 1/T, estimated from least squares refined values (□), and from area ratios of peaks in the RD-curves; *gauche* area/*anti* area (△), *anti* area/total area (●), *gauche* area/total area (○)

position, several terms are needed. Due to the limited number of terms usually included, the results in Table 4 should be taken rather as indications of trends than as quantitative findings.

The examples in Table 4 again demonstrate the effect of increasing size of substituents, reflected both by the deviation of ϕ_g from 60 ° and by the ΔH-values. (A negative ΔH means that *gauche* is the more stable conformer.)

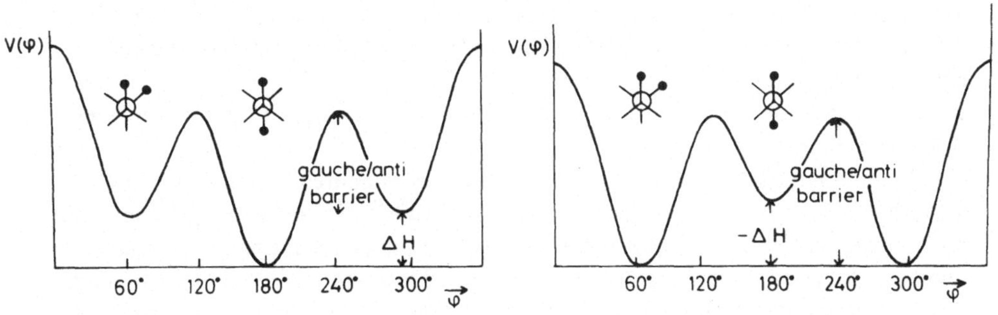

Fig. 8. Plausible torsional potential curve for 1,2-dichloroethane (left) and 1,2-difluoroethane

The experimental potential function, though not known accurately, together with the results of conformational analysis, may be used to draw conclusions about interatomic interactions within molecules. In 1,2-difluoroethane, for example, an attractive force between the fluorine atoms stabilizing the *gauche* has been estimated to be in the order of 4 kJ/mol[166, 167]. In the propanes cited in Table 4 similar attraction is observed between halogen and the methyl group. This is also the case if one of the hydrogen atoms in the methyl group has been substituted by another halogen. But under these circumstances one of the halogens turns away from the approaching group in order to avoid steric hindrance[168]. In disilane derivatives with halogen and methyl substitution the situation is different since the Si-Si-bond distance is considerably longer than the C-C-bond distance. For example, in the case of tetramethyl-1,2-dichloro-disilane[159] a *gauche* preference is observed, in spite of the fact that the *anti* conformation has four *gauche* halogen-methyl distances while the *gauche* has only two such distances. The *gauche* preference may be caused by a chlorine-chlorine attraction. It should, however, be emphasized that the conformational results in this molecule are somewhat uncertain because of the similarity of the *gauche* chlorine-chlorine distance and the corresponding chlorine-methyl distance. The barrier is probably less than 4 kJ/mol, though a freely rotating model was definitely ruled out.

The last two examples, namely $CH_2X{-}C{\equiv}C{-}CH_2X$ (X = Cl and Br) in Table 4, demonstrate that the increased separation of the rotating end groups leads to a considerable reduction in the barrier. As is to be expected, the RD-curves of these molecules are quite similar to that of hexafluorobutyne-2 (Fig. 5). The RD-curves demonstrate that a considerable proportion of the gas molecules must assume a form corresponding to the maximum areas of the potential curve which is indicative of a high degree of torsional freedom. Due to the relative small distance variation during a complete torsional revolution and also due to the relative small contribution of the torsion sensitive part of the RD-curve to the total, the noice level precludes decisive, quantitative statements. Essentially free rotation is compatible with the experimental data, as the estimated Fourier coefficients are not conclusively different from zero, though the actual estimates indicate a torsional potential with a minimum rather at *anti* than near *syn.*

6.2 1,3-Butadiene and Analogs

1,3-butadiene has been the subject of several conformational studies, by electron diffraction[169, 179] as well as by other methods[170–174]. The planar *anti*-form predominates, and to date no conclusive evidence of a second conformer has been given, though both *syn* and distorted *gauche* have been suggested for a possible additional conformer.

A rather large barrier (20 kJ/mol)[174] has been estimated between *anti* and a possible second conformer. A recent theoretical study[175] leads to an additonal, but much lower barrier of 2 kJ/mol at the *syn* position, thus separating the two *gauche* minima, estimated to be at about 40 ° from *syn*. This potential would lead to a high degree of flexibility in the *syn* region. A qualitative representation of the total potential based upon the theoretical values[175, 176], is given in Fig. 9.

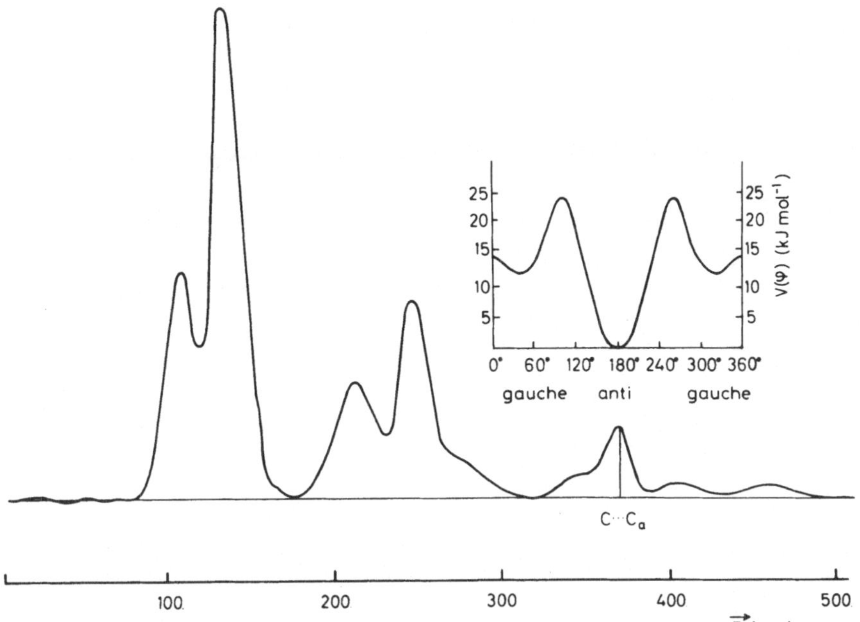

Fig. 9. Radical distribution curve and torsional potential for 1,3-butadiene

In the search for a presumed, but not conclusively detected conformer, the natural approach is to use a high a temperature as possible in the electron-diffraction experiment. Several such attempts have been made, so far in vain[169d, e]. The negative results in the case of 1,3-butadiene are not in contrast with the theroretical findings, since no rigid second conformer is to be expected.

Another approach has also been tried for 1,3-butadiene based upon study of the area under the *anti*-peak in the RD-curve compared to the rest of the RD-curve area. This is complicated by the uncertainty of the position of the zero-line in the RD-curve. The zero-line level is determined by the very inner and unobservable part of the intensity curve.

The thermal asymmetry of the $C_1 \dots C_4$ *anti*-peak gives information about the torsional motion near *anti* and demonstrates that the potential well is wider in butadiene than for torsional motion about sp^3-sp^3 linkages.

In butadiene analogs containing C=O[177–182] or N=C[183] bonds, two conformers coexist (Table 5). CH_2=N-N=CH_2[183] and O=CX-CX=O[180, 181], (X = Cl, Br), represent the rather few cases of intermediate size asymmetric barriers where the torsional potential has been determined from electron-diffraction data. In these investigations the mixture of two conformers has been determined at different temperatures. The *gauche* conformer is described as a rigid, staggered model.

Because of the observed asymmetry in the *anti*-peak, Eqs. (13) and (14) were applied over a region corresponding to $\pm 2\sigma_\phi$ in order to describe the large, although still assumed harmonic, torsional amplitude. From the determined values of ΔH, ϕ_g, and the curvature of the potential about the *anti* position, the torsional potential was estimated including V_1-, V_2-, V_3-terms in the Fourier series expansion. The resulting

Table 5. Some torsional dependent observed quantities for butadiene and analogs. Central distance (R) and u-values in pm, torsional angle (ϕ_g) and amplitude (σ_ϕ) in °. Enthalpy difference (ΔH) and Fourier terms (V_n) in kJ/mol

	ED R	ϕ_g (σ_ϕ)	u_g	u_a	ΔH	(V_n)	Stable conformer : ratio	Ref.
$CH_2=N-N=CH_2$	141.8	49.8 (20.3)	14.0[a]	6.0	5.0	V_1 = 4.60 V_2 = −2.09 V_3 = 3.05	*anti/gauche*: 78.5/21.5	183)
$CH(CH_3)=N-N=CH(CH_3)$ *trans/trans*	143.7	60(20)	9.3	6.2			*anti/gauche*: 80/20	414)
$CH_2=CH-CH=CH_2$	146.5	(16.9)		6.0			*anti*	169)
$CH_2=CH-C(CH_3)=CH_2$	147.0	(40.?)	10.	8.			*anti*	184)
$CH_2=C(CH_3)-C(CH_3)=CH_2$	150.4						*anti*	185)
$CH(CH_3)=C(CH_3)-C(CH_3)=CH(CH_3)$								
cis/cis	152.1	(26.6)		7.4			*anti*	186)
cis/trans	152.8	65.7	9.8				*gauche*	186)
trans/trans	152.1	66.7	9.8				*gauche*	186)
$CF_2=CF-CF=CF_2$	148.8	47.4	11.0[a]				*gauche*	187)
			7.					
$CCl_2=CCl-CCl=CCl_2$	148.3	78.1	8.[a]				*gauche*	188)
$CH_2=CH-CH=O$	148.4		15.	6.5			*anti*	179)
$CH_2=CH-CCl=O$		32.					*anti/gauche*: 58/42	177)
$CH_2=CBr-CH=O$		0–20.					*anti/syn*: 58/42	177)
$CH_2=CCl-CCl=O$		0–30.		6.0			*anti/syn*: 67.5/32.5	178)
$O=CH-CH=O$	152.6			6.0			*anti*	179)
$O=CCl-CCl=O$	153.4	55. (22.1)	18.0	7.2	5.8	V_1 = 4.18 V_2 = −3.76 V_3 = 4.18	*anti/gauche*: 67.6/32.4	180)
$O=CBr-CBr=O$	154.6	65.9 (30.7)	29.0	6.9	2.6	V_1 = 2.51 V_2 = −0.84 V_3 = 1.67	*anti/gauche*: 48.0/52.0	181)

$O{=}C(CH_3){-}C(CH_3){=}O$	150.7	(24.0)	*anti*	182)
$NC_2H_5{-}NC_2H_5$	147.6		*anti*	189)
$C_3H_5{-}C_3H_5$	149.9	48.7 (80)[c]	*anti/gauche*: 47.5/52.5	190)
$C_3H_4Br{-}C_3H_4Br$				
a-*trans, trans*	149.	58.8 (20.)	*anti/gauche*: 33/67	191)
a-*cis, cis*	147.	(18.8)	*anti*	192)
$C_2H_3O{-}C_2H_3O$	152.1		*anti*	193)
$C_3H_5{-}CH{=}CH_2$	147.5	60–70	*anti/gauche*: 75/25	194)
$C_3H_5{-}CH{=}O$	150.7[b]	0.	*anti/syn*: 45/55	195)
$C_3H_5{-}CCl{=}O$	151.[b]	0.	*anti/syn*: 15/85	196)
$C_3H_5{-}C(CH_3){=}O$	150.6	0.	*anti/syn*: 20/80	196)

a $u(C \cdot\cdot C)_g$.
b Average C–C distance.

c ϕ_g is defined as

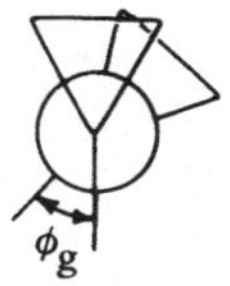

potential function gives *gauche* to *anti* barriers in the range of 0.5 – 2 kJ/mol, and the observed high u_g is consistent with such a low barrier. But, as the authors also indicate[181], even a 2 kJ/mol barrier is a very low barrier if the concept of a well defined *gauche* conformer is to be retained.

The inclusion of only three terms may be insufficient, since in the related molecule butadiene four terms in the Fourier series are found to be necessary to describe appropriately the observed frequencies[172].

A reasonable check of the reliability of the three-term Fourier potential for the butadiene analogs, would have been to calculate a RD-curve by introducing a large amplitude model using the obtained potential to determine $P(\phi)$ for the whole ϕ-interval.

In Table 5 cyclopropyl derivatives have been included for comparison. The analogy is based upon the bent bond model[60] for the double bond. *Anti* predominance is found both in cyclopropylethylene[194] and in bicyclopropyl[190]. The torsional amplitudes are dramatically increased compared to butadiene. In bicyclopropyl $\sigma_\phi = 80°$, which implies a nearly free torsional motion over a large angle region about *anti*. Figures 9 and 10 give the estimated torsional potentials together with the RD-curves for butadiene and bicyclopropyl, respectively.

For bicyclopropyl each geometric species contains two torsion dependent C . . . C-distances. The positions of these distances are given in Fig. 10 both for *anti* and *gauche*. The effect of the large torsional amplitude in bicyclopropyl is demonstrated by the broad torsional dependent area in the RD-curve without well defined peaks. This is in contrast to the well defined *anti*-peak in the corresponding part of the RD-curve for 1,3-butadiene.

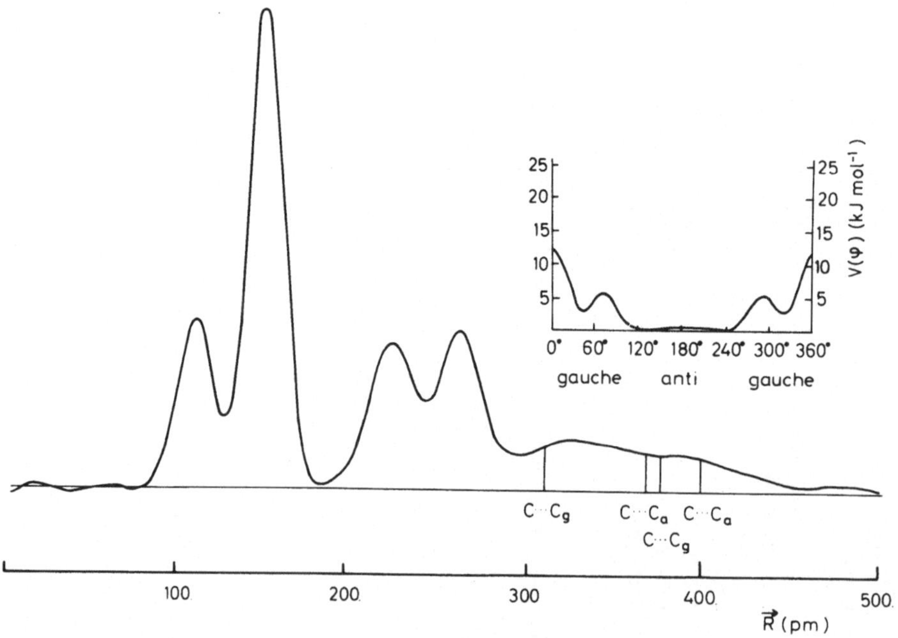

Fig. 10. Radial distribution curve and torsional potential for bicyclopropyl

The *anti*-form thus seems generally to predominate for the molecules in Table 5 if no extra strain caused by substituents has been introduced. If substituents are introduced, for example in butadiene, the *gauche*-conformer may prevail. In other cases *anti/gauche*-equilibria may occur.

6.3 Molecules with Several Torsional Degrees of Freedom

When the number of torsional degrees of freedom is increased, the intramolecular motion in gaseous molecules is increased as well. At the same time the theoretical treatment of the motion becomes more complex, and the problems that the electron-diffraction method has to face are more difficult to handle. The molecules of this category that have been subject to quantitative conformational analysis by electron diffraction so far, are limited to cases with two or a few degrees of freedom, though qualitative observations about large amplitude motion have been made also for considerably larger molecules.

Among the vast number of molecules with more than one axis of internal rotation, the open chain hydrocarbons represent a group of molecules expected to be rather flexible, since the hydrogen atom causes a minimum steric hindrance to the internal motion. However, hydrocarbons are not well suited for electron-diffraction study because of the low scattering power of the hydrogen atom. H . . . H-distances are reliably observed only in few cases, and information about large amplitude motion in hydrocarbons is mainly obtained through C . . . C- and C . . . H-distance contribution. For lack of knowledge of the hydrogen position, methyl groups are usually placed in a staggered position in accordance with the situation in ethane itself where the H . . . H-distances have been located[112)].

Parameters describing the hydrogen position are sometimes included in the least squares calculations. But both methyl twist angles and related vibrational amplitudes thus obtained are particularly sensitive to the assumptions made concerning the methyl torsional potential because of the low scattering power of hydrogen.

Bartell and co-workers have studied *n*-hydrocarbons up to 16 carbon atoms[119, 141, 197, 198)]. The results give information about structural and conformational properties, and also demonstrate the existence of large amplitude motion. Due to this motion, distances beyond 500 pm are not considered reliable enough to be included in the analysis of the large amplitude motion[198)].

In a conformational study of di-*n*-propyl ether appreciable contribution beyond 500 pm is included in the study[199)]. A conclusive determination of the conformational mixture was impossible, but the fact that the molecule exhibits a high degree of flexibility was demonstrated.

In order to obtain molecular systems in which the internal motion is easier to study, it is customary to introduce halogen atoms in the molecules because of the enhanced scattering power of these atoms. On the other hand, the larger halogen atoms restrict the internal motion more than is the case in unsubstituted molecules. Halogen substitution thus leads to systems with less torsional freedom than the parent hydrocarbons.

As examples of halogenated hydrocarbons of this type studied by electron diffraction, 1,2,3,4-tetrabromobutanes[169a)] and a series of propanes[149, 151–154, 200–204)] may be mentioned. The main object of such studies is to describe which conformers are present, their molecular structure, and, if possible, the relative abundance of the conformers.

In particular Stølevik's conformational study of a series of halogenated propanes has contributed to our understanding of the flexibility of molecules with two axes of internal rotation. The coexistence of as many as three conformers has been found in several instances. For these halopropanes the structures of the coexisting conformers have been determined as well as their mole fractions. The molecules are quite rigid, and from the study of the u-values the conclusion has been drawn that the torsional potentials are harmonic in the minimum regions.

A few examples of propane and butane derivatives are included in Table 4. The results obtained for such molecules are, as a whole, in accordance with what should be expected from the study of ethane derivatives.

In crowded molecules the torsional motion may be severely hampered. For a molecule like $(CF_3)_3CH$[205)] the electron-diffraction study may either be based upon a staggered model carrying out torsional oscillations, or a model of large geared motion described with appropriate coupling terms characterizing the interaction between the rotating groups. The results obtained for both models indicate rather large librations of the CF_3-groups. The staggered model approach leads to a twist of the CF_3-group of 15° away from the staggered position. The geared motion model calculations result in a σ_ϕ of 17°.

This result is in general agreement with an extensive analysis of the electron scattering of multiple rotor molecules by J. Karle[206)], who studied the effect on the intensity and the RD-curve in cases with low barrier and geared motion.

In order to study the geared motion quantitatively, the potential function to be used must contain the involved angle parameters and terms describing the coupling. As an example one of the suggested potential functions for propane is given in Eq. (21)[207)].

$$V(\phi_1, \phi_2) = \frac{1}{2} V_3 (2+\cos 3\phi_1+\cos 3\phi_2) - \frac{1}{2} V_3' [1+\cos 3(\phi_1+\phi_2)] \qquad (21)$$

From microwave studies V_3' was found to be 1.2 kJ/mol, i.e. about 1/10 of the V_3 term, in agreement with results obtained theoretically[207)]. This indicates that the interaction of the two rotors must be small in the propane case, though it is difficult to draw any general conclusion from this example.

7 Biphenyl and Related Compounds

The torsional motion about the central C–C bond of biphenyl and related compounds has been extensively studied by gas electron diffraction and by several other experimental and theoretical methods. The fact that biphenyl itself undergoes conformational changes by phase transition[169a, 208)], indicates that the barrier to internal rotation must be small. Not only biphenyl itself but also a series of its derivatives without *ortho* substituents are reported to be planar in the solid state[209–214] while the angle, ϕ, between the rings in the gas phase is found to be about 45 ° [215, 216]. The best value reported for biphenyl itself is 42 °. Recent studies have thrown doubt on the exact planarity of the biphenyl molecule in the solid state. From studies of the temperature dependence of the Raman spectrum from 15 to 75 K[217] and from electron paramagnetic resonance and electron nuclear double resonance studies at 1.9 K[218] the conclusion has been drawn that the molecule does not have the ideal planar symmetry. In a recent X-ray study at 110 K the old question has been revived about the possibility that the X-ray crystallography claim of a planar biphenyl molecule may have been obscured by statistical effects[219]. But in any case the inter ring angle of biphenyl in the solid state should be considerably smaller than in the gas phase.

The non-planar conformation in the gas phase is described as a result of a compromise between conjugation favoring planarity and steric repulsion between hydrogen atoms in 2-positions. These two effects seem to be balanced at an angle of twist of about 45 °. The increase in free energy that the molecule has to suffer rotating to the planar form, or near to it, is apparently overcome in the solid phase by favorable lattice free energy. It should be pointed out that cases are known where a large deviation from coplanarity is observed in the crystalline phase also for biphenyl derivatives without *ortho* substituents. For example 4,4'-dimethylbiphenyl is found to have an inter ring angle of 40 ° in the solid[220].

In a recent X-ray study of 4,4'-dichlorobiphenyl[221] a twist angle of 42 ° has been reported, in exact agreement with the electron-diffraction result of gaseous biphenyl. The 4,4'-dichlorobiphenyl crystal is isostructural with several other 4,4'-derivatives of biphenyl, however, with slightly smaller values for the twist angle.

Conformational changes also take place at phase transition between the crystalline and the gas phase for molecules such as 1,3,5-triphenylbenzene[222, 223] and hexaphenylbenzene[224, 225], though these molecules are for obvious steric reasons far from planar in the solid state. Hexaphenylbenzene is of particular interest. In the vapor phase the peripheral rings are found to be orthogonal to the central ring with a torsional amplitude of at least 10 ° to either side. This molecular arrangement is probably due to an entropy effect and is not the result of an energetically favorable conformation. From general experience a propeller shaped conformation ought to be energetically the most stable one, a conformation in fact found in the solid. On the other hand this conformation must be statistically unfavorable since, if only one of the six phenyl rings is brought out of order, the propeller form cannot be realized.

An approximation approach to study the torsonial amplitudes in biphenyls without *ortho* substituents using electron-diffraction data leads to rather large amplitudes[216]. The two molecules chosen were 3,3'-dibromobiphenyl and 3,5,3'5'-tetrabromobiphenyl. The u-value for the $Br_3 \ldots Br_{3'}$-distance was calculated from the electron-diffraction data. Only the larger of the two $Br_3 \ldots Br_{3'}$-distances appeared suited for the study. Since the total u-value is composed of contributions both from the framework vibration and from the torsional motion, an estimate of the framework vibration amplitude is needed in order to obtain the rotational amplitude, σ_ϕ. In order to estimate the framework vibration, 3,5,4'-tribromobiphenyl was studied.

The $Br_3 \ldots Br_{4'}$-distance in 3,5,4'-tribromobiphenyl is nearly of the same length as the larger $Br_3 \ldots Br_{3'}$-distance of the two other molecules. But as the $Br_{4'}$-atom lies on the axis of rotation, the $Br_3 \ldots Br_{4'}$-distance is independent of the angle of torsion. Consequently the corresponding u-value is due to framework vibration only. As a rough approximation the latter u-value was used as an estimate for the u-framework of the longer $Br_3 \ldots Br_{3'}$-distance both in 3,3'-dibromobiphenyl and in 3,5,3',5'-tetrabromobiphenyl. This led to a value of the σ_ϕ of 19 ° and 17 ° for the two compounds, respectively. This difference may be insignificant, but is at least in accordance with the expectation that the molecule with the larger moment of inertia around the central bond ought to have the smaller torsional amplitude. The torsional motion must be expected to be anharmonic. This is indicated by the fact that the larger $Br_3 \ldots Br_{3'}$-peak in the radial distribution curve of the two studied molecules in slightly asymmetric, while this is not the case for the $Br_3 \ldots Br_{4'}$-peak of the tribromo compound.

It should be emphasized that the σ_ϕ values obtained as described could not be used to draw real quantitative conclusions. However, the qualitative conclusion may be made that the torsional amplitude is probably so large that even the planar form may be expected to exist in the gaseous phase with a finite probability.

For biphenyl itself the energy dependence of the torsional angle has been studied by quantum mechanical methods. Studies based upon π-electron calculations, taking explicit account of steric effects[226–228] led to an energy minimum at a twist angle of 35 °–40 ° from planarity compared to the best electron-diffraction value of 42 ° for unsubstituted biphenyl[215]. Two different hydrogen-hydrogen potentials were used in these calculations, one leading to a twist angle of 35 °, the other one to a twist angle of 40 °. The barriers towards torsional motion are somewhat different in the two cases, but if the latter calculation is chosen from electron-diffraction criteria, the barrier at the planar form is higher (approximately 20 kJ/mol) than the barrier at the 90 ° form (approximately 8.4 kJ/mol). A more recent *ab initio* calculation[229] led to a smaller twist angle (32 °). In this calculation the positions of the four hydrogen atoms adjacent to the C-C-bridge bond as well as the bridge bond length were optimized for various values of the twist angle. The rotational barriers were found to be 5.0 kJ/mol and 18.8 kJ/mol at the planar and the perpendicular form, respectively, *i. e.* in this calculation the barrier to planarity is found to be the lower one. Both the theoretical calculations suggest a potential function for the torsional motion with a flat minimum leaving a large torsional amplitude in agreement with the electron-diffraction results.

In order to summarize our present knowledge as to the parameters of importance for understanding the torsional motion of biphenyls without *ortho* substituents, the following points may be made:

1. The electron-diffraction measurements reproduce the twist angle to 42° with an error estimate of about ± 5 °. This value corresponds to the angle of maximum probability. An exact location of the minimum of the potential function is not easy to derive with present knowledge, but the discrepancy between the electron-diffraction value and the one obtained by *ab initio* calculation is too large to be accepted.

2. The contribution of the *ortho*-hydrogen atoms to the deviation from planarity is clearly indicated both by the *ab initio* calculation and by X-ray study. Both

methods demonstrate a slight deformation of the *ortho*-hydrogen atoms to ease the steric strain.

3. The central C-C-bond, the axis of rotation, is both by X-ray and electron diffraction repeatedly found to be about 150 pm, while the *ab initio* calculation suggests a value of 153 pm. This discrepancy suggests that further quantum mechanical calculation would be of great interest.

4. Both *ab initio* calculations and electron-diffraction studies indicate a large torsional amplitude. An amplitude value of 15 ° to 20 ° seems at present to describe the torsional motion, but further studies are certainly desirable.

In order to obtain further information of the torsional motion about the bridge bond of biphenyls and related compounds it is natural to study the effect of *ortho* substitution. Such substitution should introduce more steric hindrance, increase the angle of twist, and reduce the torsional amplitude. The non-bonded interaction may, on the other hand, be eased by replacing CH groups in the 2-position by single atoms such as nitrogen or sulphur. For molecules like 2,2'-bipyridyl or 2,2'-bipyrimidyl where two or four *ortho*-CH groups have been replaced by nitrogen, one might expect planar conformation, arguing that the conjugation effect might overcome the milder steric strain. However, neither of these molecules exhibits a planar conformation in the gas phase. 2,2'-Bipyridyl was studied with 4,4'-bipyridyl as a reference substance[215]. 4,4'-Bipyridyl behaves like biphenyl, but with a sligthly smaller angle of twist (37 °). 2,2'-Bipyridyl seems to have a potential curve with a smaller barrier at the planar form than biphenyl itself. The study of this molecule is obscured by the fact that a non-planar arrangement calls for the possibility of two different conformations, one near *anti* and one near *syn*. A more recent study of 2,2'-bipyrimidyl[230] both by electron diffraction and by X-ray seems to indicate a behavior closer to biphenyl, namely a planar conformation in the crystal and an approximately 45 ° angle of twist in the gas phase. At the present state of the investigation no information is obtained about the flexibility of the molecule. The comparison of the hydrogen-hydrogen, hydrogen-lone pair, and lone pair-lone pair interaction is thus not free of controversies. Since the mentioned compounds and others related to them offer an excellent possibility of comparing these interactions, further combined electron diffraction and X-ray study ought to be carried out. Work along this line is already in progress in this laboratory.

In the present context two other molecules may also be compared, namely 3,3'-bithienyl[231] and 2,2'-bithienyl[232]. For the 3,3'-compound the angle of twist is about 30° and two conformers are found, the near *anti* conformer being slightly favored (about 60%). For the 2,2'-compound electron-diffraction study suggests nearly free rotation over a large angle interval. Here as in many cases when the electron-diffraction method is used, it is impossible at room temperature or above to distinguish between unhindered motion or a motion through a barrier of say 2.5 kJ/mol or smaller. Molecular orbital studies of the barrier to internal rotation[233] led to a two minima potential curve for 3,3'-bithienyl. The barrier between the minima was found to be about 4 kJ/mol, and the near *anti* conformer was also in this study found to be the energetically more preferable one. For the 2,2'-bithienyl the molecular orbital calculations indicate free rotation[233].

Table 6. Angle of twist (ϕ) and halogen-halogen distances in a series of biphenyl derivatives. Pauling's van der Waals distances are included as well as the differences between these data and the experimental ones

	Angle of twist, ϕ, E. D. gas study	X–X (pm)	Van der Waals distances (pm)[60]	Diff. (pm)
Biphenyl[215]	42 °			
2-Fluorobiphenyl[234]	49 °			
2,2'-Difluorobiphenyl[235]	60 °	285	270	–15
2,2'-Dichlorobiphenyl[235, 236]	74 °	346	360	+14
2,2'-Dibromobiphenyl[235]	75 °	362	390	+28
2,2'-Diiodobiphenyl[235]	79 °	382	430	+48
Perfluorobiphenyl[21]	70 °	312	270	–42

A systematic comparative study of the genuine *ortho*-substituted biphenyls helps to throw light on the torsional flexibility of the biphenyl molecule. In Table 6 biphenyl and a series of 2-substituted derivatives are listed with their best estimated angles of twist[234–236, 21]. Most of these studies date back to about 1950, and taking the dramatic development of the electron-diffraction method during the last years into consideration, the data should be treated with caution. The perfluorobiphenyl was studied only 10 years ago, and the 2,2'-dichloro-biphenyl was reinvestigated only a few years ago, essentially confirming the earlier findings[236].

It is interesting that the prevailing conformer in all the 2,2'-dihalobiphenyls is found to be one with the two halogen atoms on the same side (i. e. closer to the *syn* than to the *anti* position). In the recent work on 2,2'-dichlorobiphenyl already referred to, this finding is confirmed. Both in the gas phase, as studied by electron diffraction, and in the crystal, as studied by X-ray diffraction, the near *syn* form is the only conformer observed. The twist angle refined in the gas phase varies from 75 ° to 70 °, depending on various assumptions attached to the refinements. The average of the results of the four different refinements gives a twist angle of 72.6 °. The twist angle found in the solid is 3–7 ° smaller than the one found in the gas phase. The gas study was carried out at a temperature nearly 300 °C higher than that of the crystal study, and the oscillation is probably rather anharmonic due to the steric difficulties encountered at smaller twist angles. The packing of molecules in the crystal may also hamper torsional motion. (The reason for carrying out the electron diffraction study at such high temperature, was the search for a possible second conformer which presumably ought to have a ϕ-value about 135°. This search was negative).

Comparison of the four molecules biphenyl, 2-fluorobiphenyl, 2,2'-difluorobiphenyl and perfluorobiphenyl shows that the angle of twist increases in a reasonable way introducing more fluorine atoms in the 2-position (Table 6). A twist angle recently reported for crystalline 2-H-nonafluorobiphenyl[237] of 59.5 ° also fits nicely into the picture.

In the case of the four 2,2'-dihalobiphenyls, the angles of twist increases with the atomic weight of the halogen. The observed halogen-halogen distances are also listed and compared with Pauling's van der Waals distances[60].

Rough calculations of the potential function for torsion about the bridge bond have been carried out for some halobiphenyls by combining conjugation energy and van der Waals energy[21]. These calculation do not reproduce the experimental findings, on the contrary they suggest the more stable conformer for the 2,2'-dihalobiphenyls is closer to the *anti* than to the *syn* form. The result of these theoretical calculations was as a matter of fact the main reason for repeating the gas studies of 2,2'-dichlorobiphenyl.

The results of the 2,2'-dihalobiphenyls clearly demonstrate that not only repulsion but also non-bonding attraction may be decisive for conformational choice and for internal motion. The halogen-halogen attraction is strong enough to make the near *syn* conformation prevail to the extent that no other conformer is observable even at the highest temperature applied in the electron-diffraction experiment. The apparent systematic, and in some cases rather large, deviation from the London-force distance, as demonstrated by the last column of Table 6, remains to be explained. It is to be hoped that the experimental findings for the 2,2'-dihalobiphenyls will animate further theoretical studies.

Of course conformational decisive non-bonded attraction is not limited to halogen-halogen interaction. For example 2,2'-diaminobiphenyl, as studied by X-rays in the crystal, also prefers the near *syn* conformation with an angle of twist equal to 52 °[238].

The torsional motion of perfluorobiphenyl was studied using more advanced electron-diffraction technique, and both the torsional amplitude and the torsional barrier through the 90 ° position was refined[21]. The torsional amplitude (σ_ϕ) was found to be $10° \pm 3°$. It is considerably smaller than the value of 15° to 20° estimated for biphenyl derivatives without *ortho* substituents, which is a reasonable result. The barrier through the 90 ° position was found to be somewhere between 1.7 and 8.4 kJ/mol, while the barrier through the planar position was too high to be estimated by the electron-diffraction technique.

The torsional motion of biphenyl and related compounds is a typical large amplitude motion. The accumulated knowledge from a series of molecules in this group has led to a fairly good qualitative description of the motion. Unfortunately the quantitative description leaves much to be desired. Taking advantage of the improvements in the electron-diffraction method and applying suitable combinations with other methods, there are reasons to believe that this deficiency should be remedied.

8 Cyclic Compounds

8.1 Four-Membered Rings

The ring-puckering problem of four-membered rings has for many years attracted considerable interest from electron diffractionists and spectroscopists. A large body of information on this phenomenon has been gathered. A comprehensive and profound review on the dynamics and barrier-height determination of ring-puckering and pseudorotation potentials has recently been presented by Gwinn and Gaylord[239].

The early diffraction work of Dunitz and Schomaker[240] in 1952 on the prototype molecule cyclobutane revealed that the carbon ring was either planar with large amplitude out-of-plane motions or, alternatively, permanently bent. Eight years later it was found that cyclobutane was static non-planar with a ring-puckering angle of about 35 ° [241]. The same value was later found by spectroscopy[242]. A microwave study of cyclobutyl bromide[243] also led to a non-planar carbon ring but with a slightly smaller angle (29.4 °). It was then established that cyclobutane exists with distinguishable axial and equatorial hydrogens similar to cyclohexane. It was also established by electron diffraction[244] and by microwave studies[243] that the equatorial positions were energetically favored as in cyclohexane derivatives. The electron-diffration study[244] of four 1,3-dihalocyclobutanes, viz. *trans*-1,3-dibromo-, *trans*-1,3-chlorobromo-, *cis*-1,3-dibromo-, and *cis*-1,3-chlorobromo-cyclobutane led to an α angle of 33 ± 2 °. The puckering-angle values obtained from electron diffraction were determined both for cyclobutane itself and for its derivatives neglecting the shrinkage effect. When this is included, as shown in a recent work[245], a smaller angle is obtained (about 26 ± 3 °).

The study of the 1,3-dihalocyclobutanes shows that the barrier of the puckering potential in the planar form must be at least 4 kJ/mol. The halogen-halogen peaks in the radial distribution curve are well defined and well resolved as shown in the case of *cis*-1,3-dibromocyclobutane, Fig. 11. The figure also clearly demonstrates the preference of the equatorial position.

For the *trans*-1,3-chlorobromocyclobutane there is a conformational mixture since the two positions *a* and *e* may be occupied either by the bromine or by the

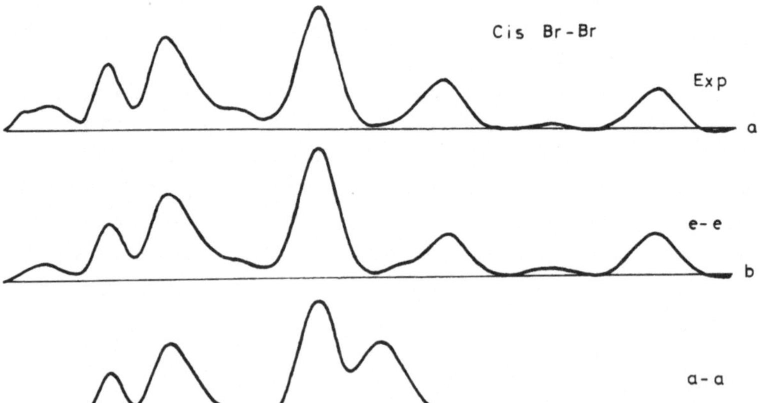

Fig. 11. Radial distribution curves for *cis*-1,3-dibromocyclobutane. Experimental RD curve (a). Theoretical RD curve for a model with the two bromine atoms in equatorial positions (b). Theoretical RD curve for a model with the bromine atoms in axial positions (c)

chlorine atom. A mixture with about 60% of the conformer with bromine in an *a*-position gives the best agreement with the electron-diffraction data.

Two further halogenated cyclobutanes have in recent years been studied by electron diffraction. Invariably, the rings are found to be non-planar. In octafluorocyclobutane[246), α was found to be 17.4° in one study and 24 ± 3° in another one[247). In 1,1-dichlorohexafluorocyclobutane[247) the puckering angle is 23.2 ± 2.5 °.

Molecules with fused carbon-ring systems are considerably less puckered than cyclobutane itself.

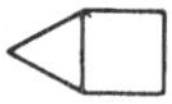
(*1*)

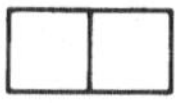
(*2*)

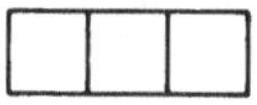
(*3*)

Bicyclo[2.1.0]pentane (*1*) seems to be planar[248): in bicyclo[2.2.0 hexane] (*2*) α is 11.5 ± 1.8 ° [249), and in the *syn* and *anti* isomers of tricyclo[4.2.0.0]^{2,5} octane (*3*) α is 8–9 ° [250).

Four-membered rings with one or more hetero atoms offer the opportunity of investigating the influence of electro-negativity, hybridization, atomic size, etc., on the puckering problem. Several such molecules have thus been investigated by electron diffraction and spectroscopy.

In 1-silacyclobutane (*4*)[251) α is 33.6 ± 2.1 °, in 1,1-dichloro-1-silacyclobutane[251b) (*5*) α is 30 ± 5 °, in 1,1,3,3-tetrachloro-1,3-disilacyclobutane (*6*)[251b) α

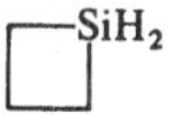

(*4*)

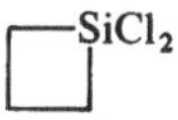

(*5*)

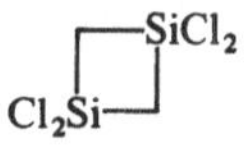

(*6*)

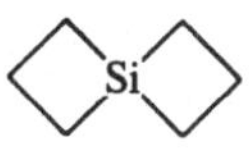

(*7*)

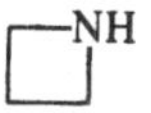

(*8*)

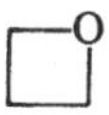

(*9*)

(*10*)

(*11*)

is reduced to 14 ± 3 °, and in 4-sila-3,3-spiroheptane (*7*) α is 30.1 ± 2.2 ° [251b). In azetidine (*8*) Mastryukov et al.[252) found that the puckering angle is 33.1 ± 2.4 °. Interestingly, oxetane (*9*) is nearly planar with a very low barrier[253) (See Table 7).

In a combined electron diffraction and spectroscopy analysis[254) the puckering angle of thietane (*10*) was determined as 26 ± 2 °. Moreover, selenetane (*11*)[255) is non-planar with α = 29.5 °.

Two molecules with two heteroatoms in the ring have recently been studied. Both tetrafluoro-1,3-dithiane (*12*)[256) and tetrafluoro-1,3-diselenetane (*13*)[251) are planar.

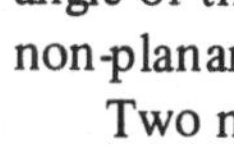

(*12*)

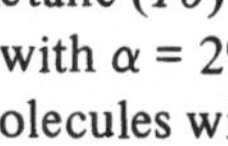

(*13*)

Table 7. Puckering angle (α) and barrier to inversion of some selected four-membered rings

Compound		Method	Ref.	Barrier (kJ/mol)
H_2C-CH_2 H_2C-CH_2	35 °	ED	[241]	6.018 ± 0.022
	26 ± 3 °	ED	[245]	
H_2C-SiH_2 H_2C-CH_2	33.6 ± 2.1 °	ED	[251]	5.26
H_2C-NH H_2C-CH_2	33.1 ± 2.4 °	ED	[252]	5.27
H_2C-O H_2C-CH_2	0 ° [a]	MW	[253]	0.1856 ± 0.0006
H_2C-S H_2C-CH_2	26 ± 2 °	ED	[254]	3.28 ± 0.02
H_2C-Se H_2C-CH_2	29.5 ± 1 °	MW	[255]	4.58 ± 0.04

[a] See text.

So far, electron diffraction has not been used for a real quantitative determination of barrier heights of four-membered rings. For reference, some barriers determined by spectroscopic methods are collected in Table 7. They have been abstracted from the compilation of Gwinn and Gaylord[239]. Typically, the barriers to puckering are fairly small. Most of them are less than 6 kJ/mol.

8.2 Five-Membered Rings

In contrast to the findings for the cyclobutanes where the large amplitude motions mainly consist of conversion between rather rigid forms, the cyclopentanes exhibit more complex conformational and dynamic properties. Pseudorotation is a prominent large amplitude motion prevailing not only in cyclopentane but also in other five-membered rings. If the barrier to pseudorotation is high, distinct conformations may exist. In this case, the envelope conformation which has maximum C_s symmetry

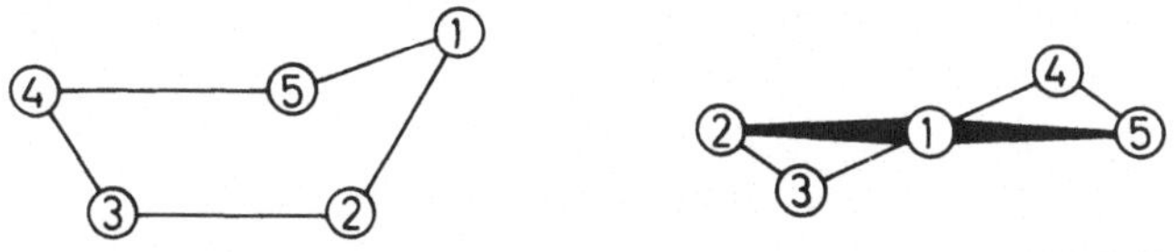

Fig. 12. Envelope (left) and half-chair conformations of five-membered rings

or the half-chair (maximum C_2 symmetry) are usually preferred. Moreover, some five-membered rings are found to be planar.

Barriers to pseudorotation have been determined mainly by various spectroscopic methods[239], while electron diffraction has provided important conformational and structural data.

Cyclopentane was studied as early as in 1931 by Wierl[258] who found that the molecule was planar. In 1946 Hassel and Viervoll[259] found that the carbon ring deviates slightly from planarity. This has later been confirmed by repeated investigations[241, 260]. Spectroscopic studies[239] have shown that this molecule undergoes nearly free pseudorotation.

In the related molecule cyclopentasilane, $(SiH_2)_5$, rapid pseudorotation has been found in a spectroscopic investigation[261]. Both the C_2 and the C_s model fit the electron-diffraction data well, but it could not be decided whether pseudorotation was static or dynamic in this study[262].

Several five-membered rings with one or more heteroatoms have been studied by electron diffraction. Tetrahydrofuran[263] (*14*) was found to have a non-planar ring.

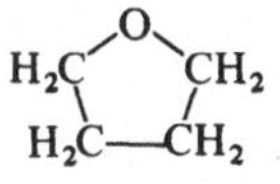

(*14*)

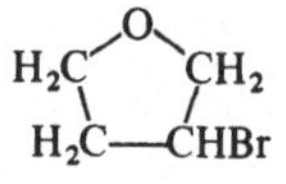

(*15*)

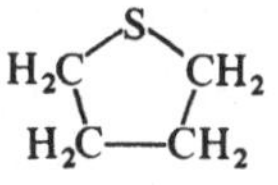

(*16*)

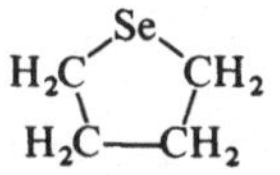

(*17*)

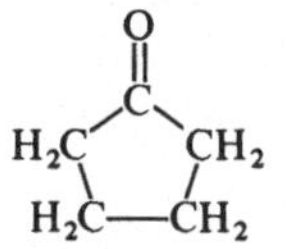

(*18*)

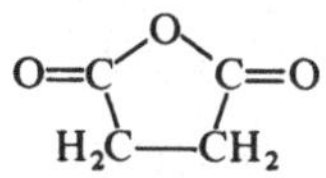

(*19*)

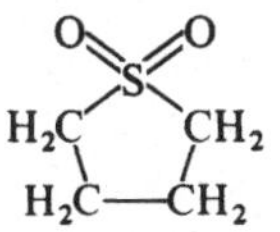

(*20*)

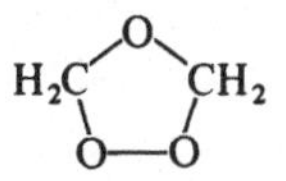

(*21*)

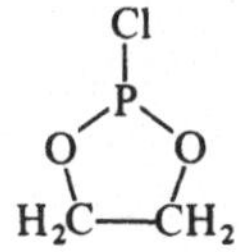

(*22*)

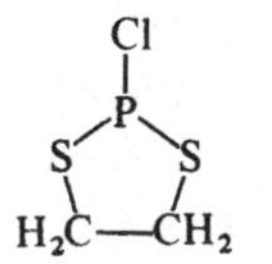

(*23*)

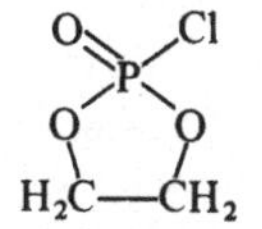

(*24*)

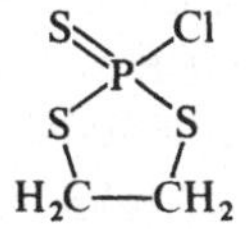

(*25*)

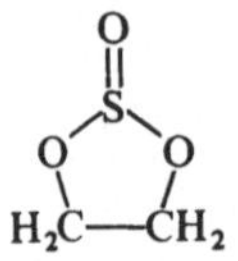
(26)

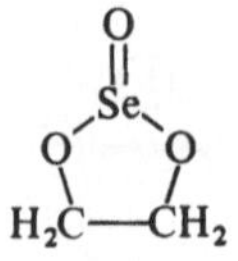
(27)

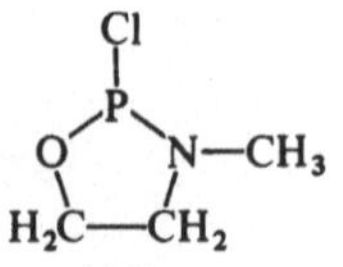
(28)

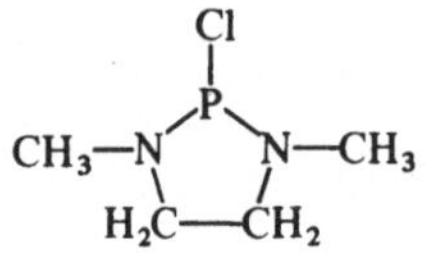
(29)

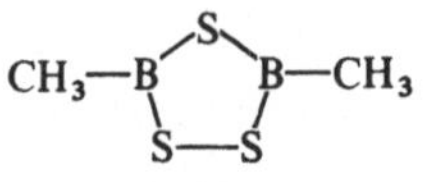
(30)

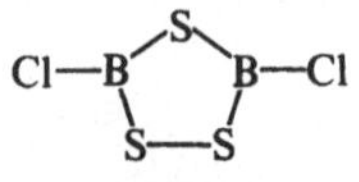
(31)

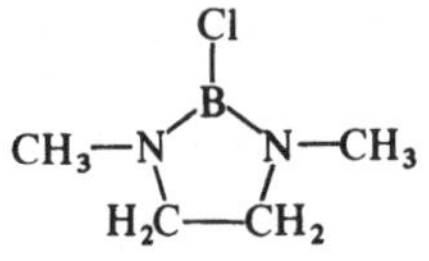
(32)

Evidence for essentially free pseudorotation was encountered in keeping with spectroscopic results[239]. As expected, almost free pseudorotation exists in 3-bromotetrahydrofuran (*15*)[264].

Tetrahydrothiophene (*16*) prefers the C_2 conformation[265]. It was impossible in this case to obtain a good fit between the experimental and the theoretical electron-diffraction data by assuming C_s symmetry. The same thing was found for tetrahydroselenophene (*17*) by electron diffraction[266] and by microwave spectroscopy[267]. The pseudorotational barriers of tetrahydrothiophene and tetrahydroselenophene must thus be considerable larger than in tetrahydrofuran since the two first-mentioned molecules exist in well-defined conformations. This is reminiscent of the puckering potential found for four-membered rings where a very low barrier was determined for oxetane while thietane and selenetane have more "normal" potentials (see Table 7).

The C_2 conformation has been found for cyclopentanone (*18*)[268], while succinic anhydride (*19*) is planar[269]. However, in tetramethylsuccinic anhydride[270] and tetrafluorosuccinic anhydride[271] conformations with non-planar rings were observed.

Both the C_s and C_2 models fit well for tetramethylene sulfone (*20*)[272]. 1,2,4-trioxacyclopentane (*21*) has been studied by electron-diffraction[273] and microwave spectroscopy and found to exist in the C_2 conformation[274].

Ethylene chlorophosphite[275] (*22*) and 2-chloro-1,3-dithia-2-phospholane (*23*)[276] both seem to prefer the C_s conformation with axial P-Cl bonds. On the other hand, ethylene chlorophosphate (*24*) and ethylene chlorotrithiophosphate (*25*) have their rings in the half-chair conformation[277]. Ethylene sulfite (*26*) and ethylene selenite (*27*) have planar or almost planar rings[278]. However, in 1,2-dimethylethylene sulfite a non-planar ring best fits the data[279]. N-methyl-2-chloro-1,3,2-oxapholane (*28*) has an envelope conformation with valence angles of nitrogen almost coplanar[280]. For N,N-dimethyl-2-chloro-1,3,2-diazaphospholane (*29*) the C_s conformation with equatorial methyl groups and axial P-Cl bond fits the data well[281]. The C_2 model did not yield better agreement[281].

Dimethyl-1,2,4-trithia-3,5-diborolane (*30*)[282] and dichloro-1,2,4-trithia-3,5-diborolane (*31*)[283] are at least approximately planar as is 1,3-dimethyl-2-chloro-1,3-diaza-2-boracyclopentane (*32*)[284].

As shown above, five-membered rings with only single bonds within the ring indeed exhibit a varied dynamical and conformational behavior. Seip and co-workers[285] have carried out molecular mechanics computations for several of these systems and generally obtained good agreement with the experimental results. It seems that this kind of calculation may become a very helpful tool for the study of large amplitude vibrations of rings in general.

Five-membered rings with one double bond are normally found to be in an envelope conformation, or they are planar. For the non-planar molecules a puckering potential will exist. The barrier to puckering is generally quite low. For example, in cyclopentene a barrier of about 2.4 kJ/mol has been determined[239].

Electron-diffraction studies have been made for several of these compounds. The puckering angle α of cyclopentene (*33*) was determined as 29.0 ± 2.5 °[286]; this angle is about 8 ° smaller in perfluorocyclopentane (*34*)[287], viz. 21.9 ± 0.5°. In

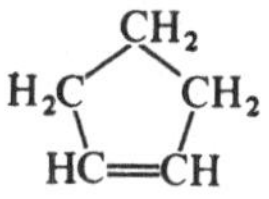
(*33*)

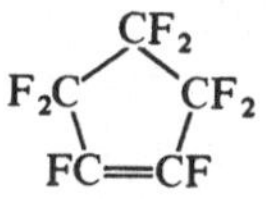
(*34*)

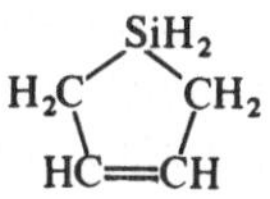
(*35*)

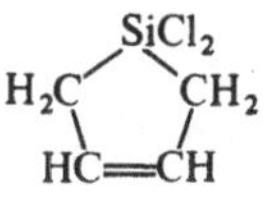
(*36*)

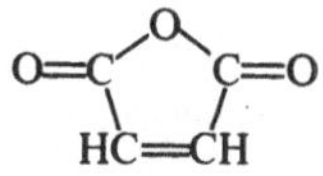
(*37*)

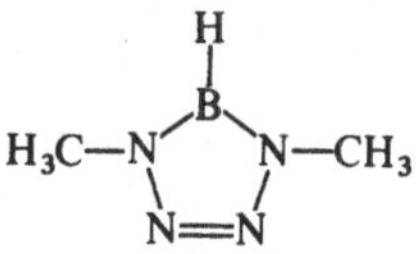
(*38*)

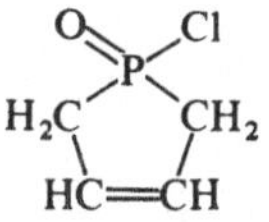
(*39*)

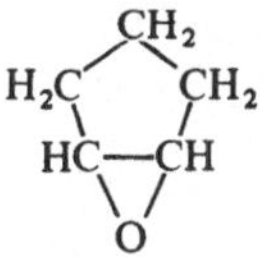
(*40*)

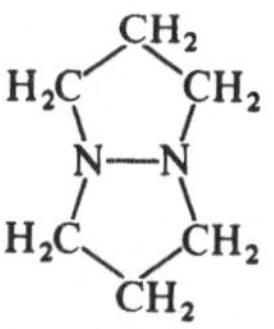
(*41*)

1-silacyclopent-3-ene (*35*)[288] α is 15.7 ± 7.7 °. 1,1-Dichloro-1-silacyclopent-3-ene[289] (*36*) is also non-planar with α = 16.8 ± 3.1 °. Maleic anhydride (*37*)[290] and dimethylcyclotetrazenoborane (*38*)[291] are planar or very nearly so. 1-Oxo-1-chloro-phosphacyclopent-3-ene (*39*)[292] prefers the envelope conformation with the P=O bond *cis* to the C=C double bond.

Cyclopentene oxide (*40*)[293] and 1,5-diazabicyclo[3.3.0]octane (*41*)[294] are similar to the cyclopentene derivatives in that rotation about one bond is very much restricted. (*40*) takes a boat conformation[293], while the rings are twisted in (*41*)[294].

8.3 Six-Membered Rings. Cyclohexane and Its Derivatives

The studies of cyclohexane and its derivatives by Hassel and co-workers in the late thirties and early forties using mainly the electron diffraction method laid the foundation of conformational analysis. In 1943 Hassel[295)] summarized that cyclohexane exists mainly in the chair conformation as distinct from any other possible conformation. The chair conformation will have distinguishable axial, *a*, and equatorial, *e*, substituents. (See Fig. 13). The equatorial position is the energetically favored one. Furthermore, Hassel stated that there is a rapid inversion of the ring with an associated low barrier. This motion interchanges the *a* and *e* positions with the result that *a* and *e* conformers cannot be isolated.

The structure of cyclohexane itself[296)] as well as the barrier and the dynamics of the ring inversion have been the objects of several studies in the sixties and seventies. Anet and Bourn[297)] in a NMR-study of $C_6D_{11}H$ found the following thermodynamic activation parameters: $\Delta H^{\ddagger} = 45.6$ kJ/mol, and $\Delta S^{\ddagger} = 12.1$ J/mol K. The activation enthalpy of 45.6 kJ/mol is probably fairly close to the barrier height. This value is in good agreement with very recent molecular mechanics calculations[298)]. The exact inversion path is unknown, and is quite likely rather complicated. This is indicated by the molecular mechanics calculations in which it was found that several conformations have rather similar energies. E. g., the D_2 twist and the C_2 twist conformation were both computed to be 22 kJ/mol less stable than the chair, while the C_{2v} boat and the C_2 boat conformations were calculated to have slightly higher energies, namely 28 kJ/mol as compared to the stable chair. All these four geometrical forms considered by Allinger et al.[298)] may be local minima on the inversion potential surface. Further minima may also exist. There is experimental evidence obtained from substituted cyclohexanes that twisted conformations or boat forms are about 20–25 kJ/mol less stable than the chair[299)], in good agreement with the molecular mechanics calculations[298)].

Ab initio calculations are not very convincing as the boat conformation was computed to be 61 kJ/mol and the half-chair 68 kJ/mol less stable than the chair[300)]. This is about three times the expected values (20–25 kJ/mol).

Mono- and 1,1-disubstituted cyclohexanes. There are several electron diffraction investigations of mono-substituted cyclohexanes. Cyclohexylfluoride[301)] exists as 57% *e* and 43% *a* which means that the equatorial position is favored by about 710 J/mol. A microwave study[302)] yielded 1.6 ± 1.2 kJ/mol for this energy difference. Cyclohexylchloride[303)] is very similar to the fluorine derivative in that the gas phase

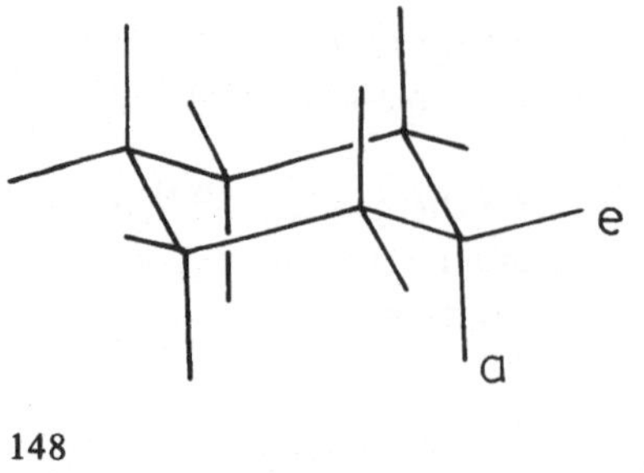

Fig. 13. Chair conformation of cyclohexane with equational (e) and axial (a) positions indicated

is composed of 45% axial and 55% equatorial conformers. Methylcyclohexane[304)] was found to have the methyl group only in the equatorial position. If this group were placed axially, the distance between it and the axial hydrogen in the 3 position would be rather short and presumably repulsive. This 1,3-interaction, which is usually repulsive, is important for conformational preferences, a fact pointed out by Hassel[295)].

In 1,1-dimethylcyclohexane one of the methyl groups must of course be in the axial position and thus experience repulsion from the hydrogens in the 3-position. In this case, a flattening of the ring was observed[305a)].

2,2,6-Trimethylcyclohexanone[306)] has a carbonyl group instead of one axial and one equatorial hydrogen atom. This molecule too takes a chair conformation.

1,2-disubstituted cyclohexanes. One consequence of cyclohexane ring conversion is that *trans*-1,2-disubstituted cyclohexanes with identical substituents can exist as *ee* or *aa* conformers as shown in Fig. 14 while the corresponding *cis ae* conformers only can exist as rapidly interconverting optical antipodes. Indeed, *trans*-1,2-dibromocyclohexane was the first molecule for which the coexistence in the gas phase of two conformations was proved experimentally[5)]. The electron-diffraction study indicated that the molar ratio of *ee* to *aa* was about 1.5, suggesting that *ee* is more stable by 0.8–1.2 kJ/mol[6)].

Trans and *cis*-decalin[307)] take the conformations shown in Fig. 15. Inversion cannot take place in *trans*-decalin, while *cis*-decalin inverts into its optical antipode if both rings invert simultaneously.

Several further molecules which may be regarded as 1,2-disubstituted cyclohexanes have been studied by electron diffraction. Principal conformational findings for several of these compounds are summarized in Table 8.

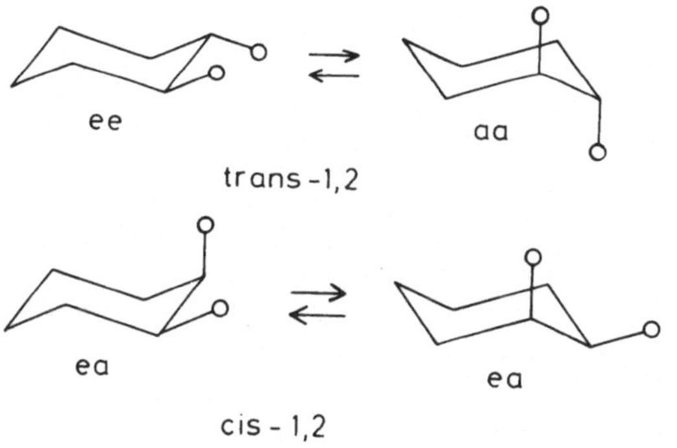

Fig. 14. Conformational possibilities for *trans*-and *cis*-1,2-disubstituted cyclohexanes

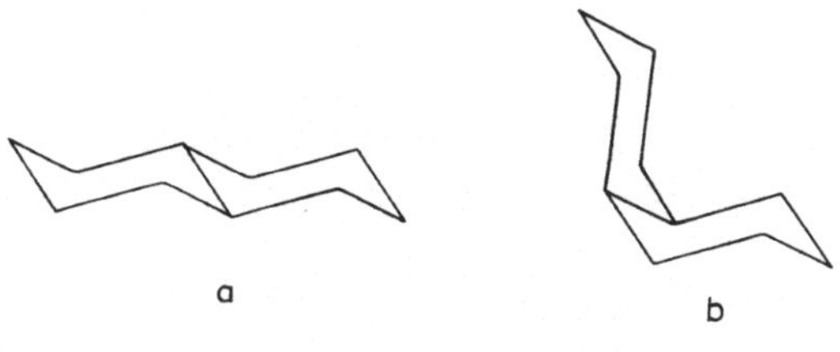

Fig. 15. *Trans*-**(a)** and *cis*-**(b)** decalin

Table 8. Selected six-membered rings studied by electron diffraction

Compound	Conformation	Remarks	Ref.
Bicyclo[4.1.0]heptane (*42*)	Half chair		308)
7,7-Dichlorobicyclo[4.1.0]-heptane (*43*)	Half chair		309)
Cyclohexene epoxide (*44*)	Half chair		310)
cis- and *trans*-Tricyclo-[5.1.0.0^{2,4}]octane (*45*)	Distorted chair	Two conformers for *trans*	311)
8,8-Dichloro-1,4,4-trimethyl-tricyclo[5.1.0.0^{3,5}]octane (*46*)	Presumably planar central ring		312)
cis- and *trans*-Bicyclo[4.2.0]-octane (*47*)	Distorted chair		313)
trans-2-Decalone (*48*)	Indications that both rings are distorted chairs		314)
10-Methyl-*trans*-2-decalone (*49*)	Both rings chair		315)
1,1-Dimethyl-*trans*-2-decalone (*50*)	Distorted chair		316)
Perhydroantracenes (*51*)	Chair		317)

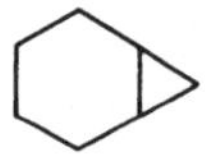

(*42*)

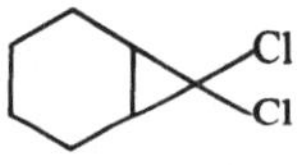

(*43*)

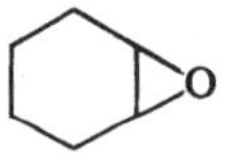

(*44*)

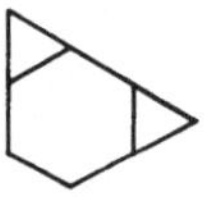

(*45*)

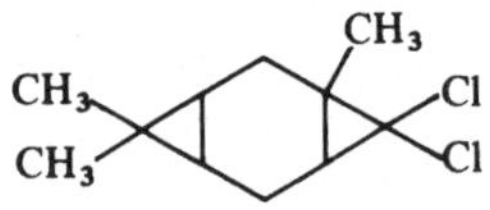

(*46*)

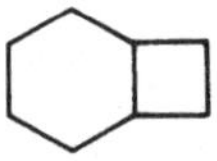

(*47*)

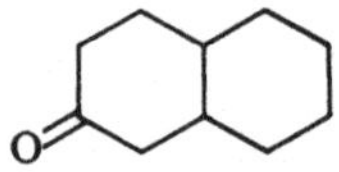

(*48*)

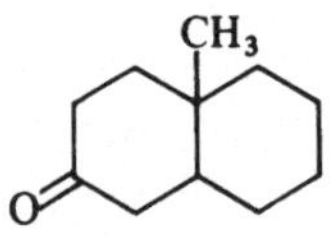

(*49*)

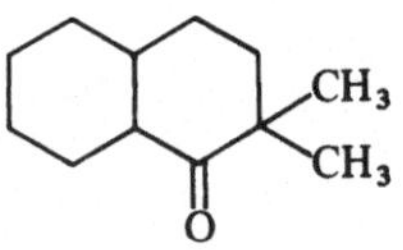

(*50*)

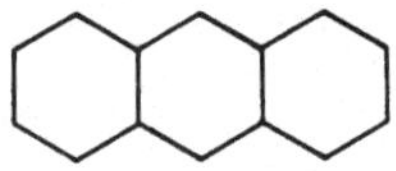

(*51*)

1,3-disubstituted cyclohexanes. Very few 1,3-disubstituted cyclohexanes have been studied by electron diffraction. β-Pinene (Fig. 16) may be regarded as a 1,3-disubstituted cyclohexane. This molecule contains two fused six-membered rings and one of them must be in a boat-like conformation. Naumov and Bezzubov[318)] found

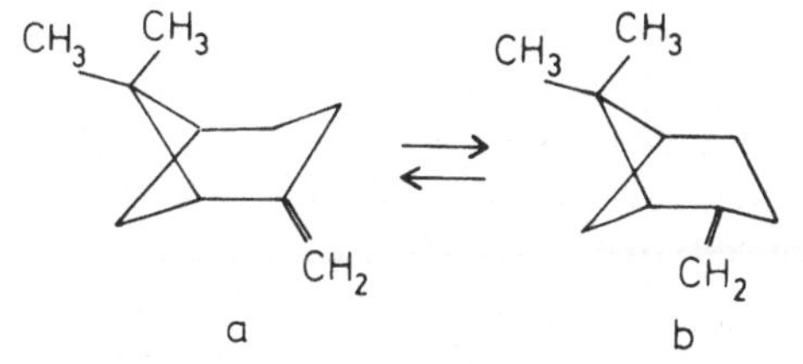

Fig. 16. Boat **(a)** and chair **(b)** conformations of β-pinene

that this molecule exists as 65% in the "chair" and 35% in the "boat" conformation of Fig. 16.

1,4-disubstituted cyclohexanes. Both the *cis* and the *trans* 1,4-disubstituted cyclohexanes will exist as *meso* compounds if the two substituents are identical. The *cis*-isomer can only have the *ae* substituent configuration, while the *aa* and *ee* conformers are possible for the *trans*-isomer. *Trans*-1,4-dichloro[319] and *trans*-1,4-dibromocyclohexane[319] have been studied by electron diffraction, and the amounts of the *aa* and *ee* conformations were found to be practically equal for both molecules. Hence, the energy differences were estimated to be less than 0.7 kJ/mol.

It has been pointed out[6] that *trans*-1,4-dihalocyclohexanes are ideally suited for determination of changes in free energy, enthalpy, and entropy by electron diffraction, since the ratio between the conformers may be found quite accurately by comparing areas of the appropriate peaks in the experimental RD curve (see Fig. 17). Accurate thermodynamic parameters of gas phase conformational equilibria are rather scarce and these molecules may perhaps be able to provide some much-wanted high-quality data with the aid of the modern electron-diffraction technique.

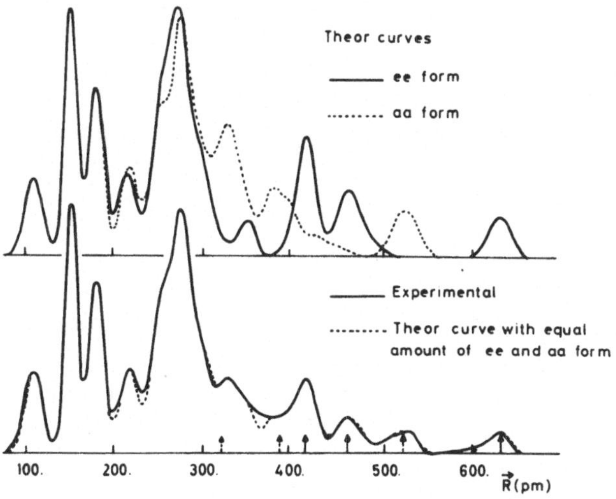

Fig. 17. Theoretical and experimental RD curves of *trans*-1,4-dichlorocyclohexane

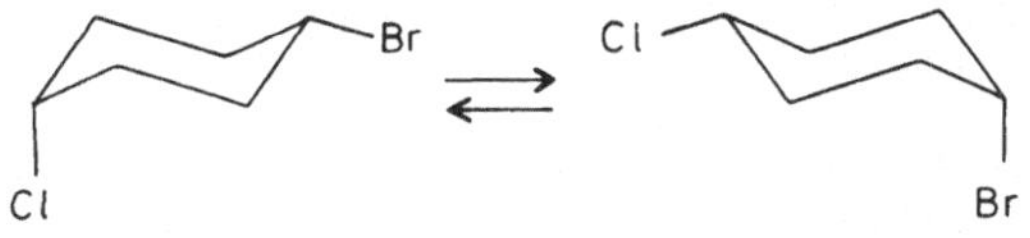

Fig. 18. Conformational equilibrium of *cis*-1,4-chlorobromocyclohexane

Cis-1,4-chlorobromocyclohexane[320)] is representative for a molecule with non-identical substituents. In this case, conformations with a *a*-chloro- and *e*-bromo substituents, or *vice versa* are possible. It was found that this compound exists almost completely in *e*-chloro-*a*-bromo conformation. This contrasts with the results of the corresponding cyclobutane derivatives earlier referred to (p. 142).

cis-1,4-Ditertiary butylcyclohexane[321)] will have one tertiary butyl group in the *a* position. This will be sterically very unfavorable because of the size of the tertiary butyl group. The electron diffraction data strongly indicates that the ring is distorted[321a)]. A composition of 1/3 chair and 2/3 non-chair ring conformation yielded the best fit to the data.

In both *cis*-and *trans*-4-tertiary butyl-1-chlorocyclohexane[305b)] the bulky tertiary butyl group will be equatorial. In these two molecules only a small deviation from the ideal cyclohexane geometry was found for their rings.

Polysubstituted cyclohexanes. If there is only one type of substituent, the conformation with a maximum number in the equatorial positions is generally favored. If several different substituents are attached to the cyclohexane ring, it is not always obvious which position is the more stable.

Some polysubstituted, mainly halogenated cyclohexanes have been studied by electron diffraction. Two conformers are possible for 1,2-dichloro-4,5-dibromocyclohexane[322)] (Fig. 19). The *aa*-chloro-*ee*-bromo conformer predominates in the gas phase. Deviation from idealized geometry was seen in this case. The *a*-C-Cl bonds are bent away from the principal axis of the ring by approximately 8 °. The *e*-C-Br bonds are also bent away from each other by about 3 °. A similar finding was made for dodecafluorocyclohexane, C_6F_{12},[323)] where the axial fluorine atoms are bent away 6.2 ° from the principal axis of the ring. This kind of deviation from cyclohexane geometry is quite common and was also found for 1,2,4,5-tetrachlorocyclohexane in the crystalline state[324)].

Five of the eight theoretically possible 1,2,3,4,5,6-hexachlorocyclohexanes have been studied and their preferred gas phase conformations determined[325)].

Cyclohexene and derivatives. The stable conformation of cyclohexene is the half-chair[326)] (Fig. 20).

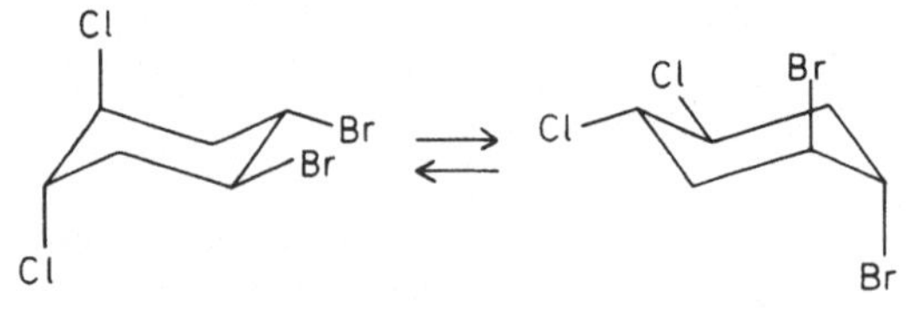

Fig. 19. Conformational equilibrium of 1,2-dichloro-4,5-dibromocyclohexane

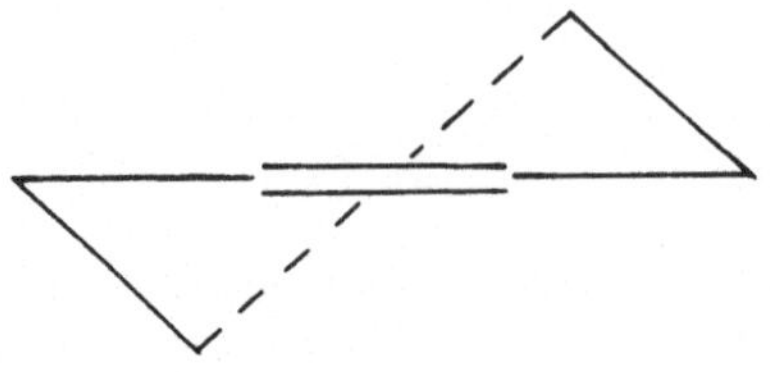

Fig. 20. The half-chair conformation of cyclohexane

If substituents are placed in the 3 and 4 positions, axial and equatorial conformations may arise. In 3-chlorocyclohexene (*52*)[327] 80% is axial and 20% equatorial,

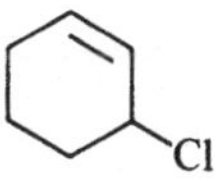

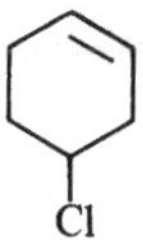

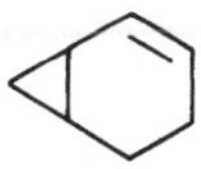

(*52*) (*53*) (*54*) (*55*) (*56*)

Table 9. Selected six-membered rings with hetero-atoms studied by electron diffraction

Compound	Conformation	Remarks	Ref.
1,3-Dioxane (*57*)	Chair		332)
1,4-Dioxane (*58*)	Chair		333)
1,3,5-Trioxane (*59*)	Chair		334)
2,4,6-Trimethyltrioxane (paraldehyde) (*60*)	Chair	Methyl groups equatorial	335)
1,3-Dithiane (*61*)	Chair		336)
1,4-Thioxane (*62*)	Chair		337)
4-Thiacyclohexanone (*63*)	Chair		338)
Piperazine (*64*)	Chair		333, 339)
N,N-Dimethylpiperazine (*65*)	Chair	Methyl groups equatorial	333)
2,2,6,6-Tetramethyl-4-piperidinone-1-oxyl free radical (*66*)	Chair		340)
2,2,6,6-Tetramethyl-hydroxyl-4-piperidinone (*67*)	C_s suggested	Considerably distorted chair	341)
Trimethylene sulfite (*68*)	Chair	S=O axial	342)
trans-4,6-Dimethyl trimethylene sulfite (*69*)	Probably chair		343)
Trimethylene selenite (*70*)	Chair	S=O axial	344)
Trimethylene chlorophosphite (*71*)	Chair	Axial P–Cl predominates	345)
1,3-Dimethyl-2-chloro-diazbora-cyclohexane (*72*)	C_s	C_5 out of plane formed by the other atoms	346)
Thiantien (*73*)		Dihedral angle 128–130 °. Inversion barrier larger than 4 kcal/mole	347)

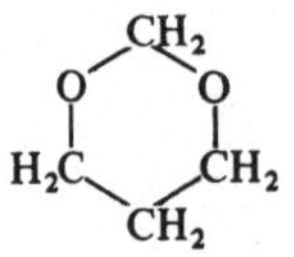
(*57*)

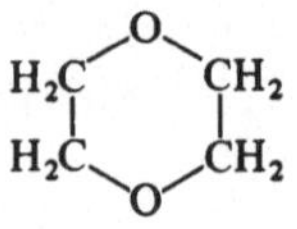
(*58*)

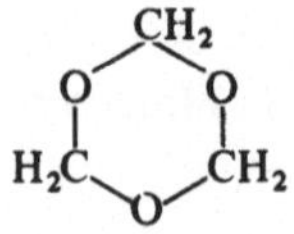
(*59*)

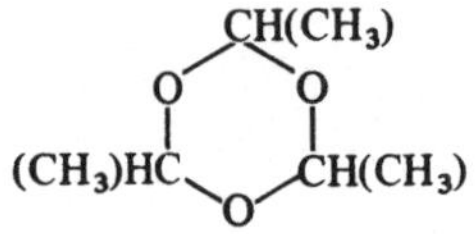
(*60*)

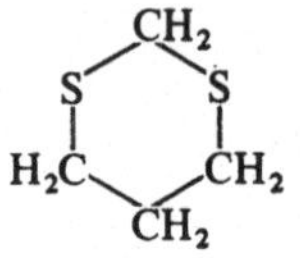
(*61*)

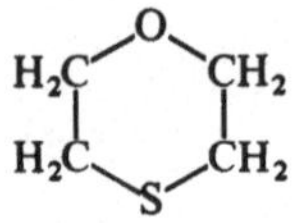
(*62*)

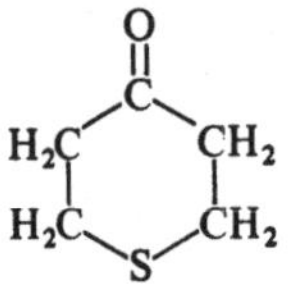
(*63*)

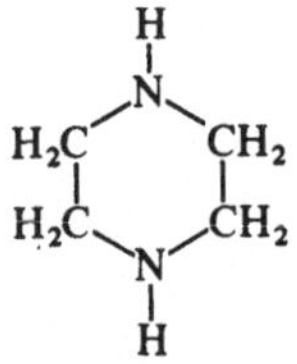
(*64*)

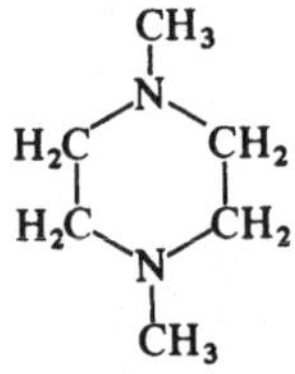
(*65*)

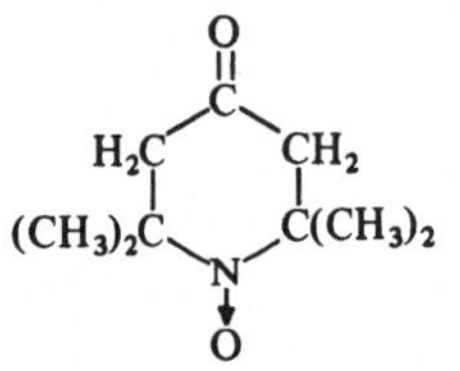
(*66*)

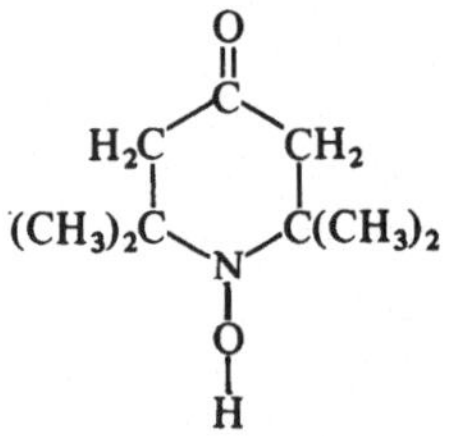
(*67*)

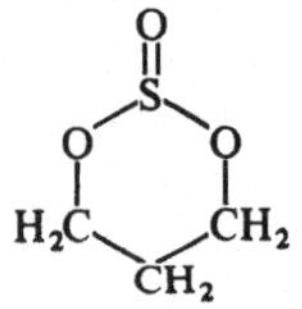
(*68*)

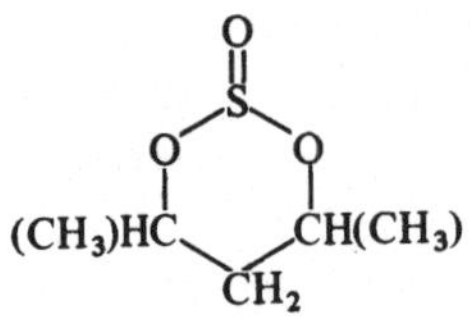
(*69*)

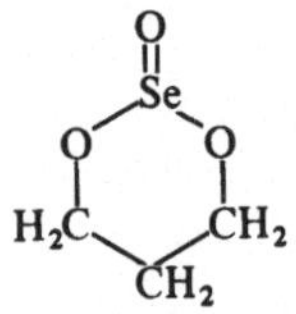
(*70*)

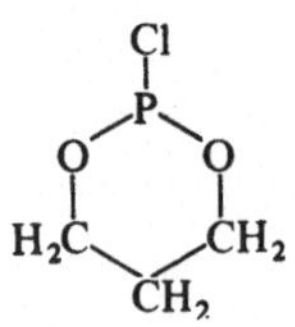
(*71*)

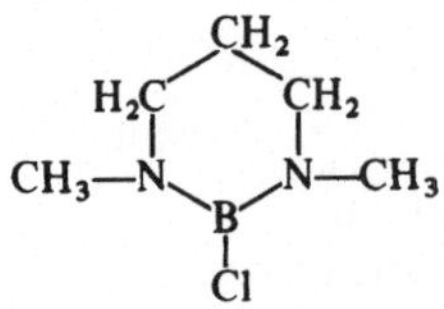
(*72*)

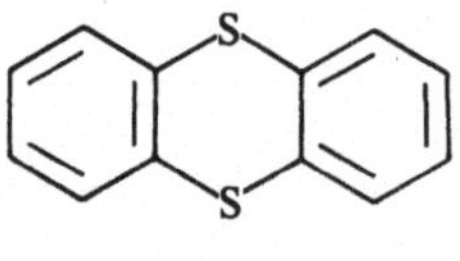
(*73*)

while 40% takes the axial and 60% the equatorial position in 4-chlorocyclohexane (*53*)[327]. Five isomers of 3,4,5,6-tetrachlorocyclohexene-1 have been studied[328] and their conformational preferences determined. In the β-isomer, evidence for the presence of two conformers was found. 1,3-Cyclohexadiene (*54*) is twisted 18.1 ± 1.0 ° about the single bond connecting the two double bonds[329]. 1,4-Cyclohexadiene (*55*) is planar at equilibrium[330]. In bicyclo[4.1.0]-2-heptene (*56*) a 30% to 70% mixture of two twisted conformations was taken as the best model at room temperature[331].

Cyclohexane derivatives with hetero atoms. Electron diffraction studies have been made for many molecules with hetero atoms. These molecules are analogous to cyclohexane derivatives. Principal findings pertaining to large amplitude vibration are summarized in Table 9.

It is seen from this table that the chair conformation is preferred when one or more ring methylene groups are substituted by nitrogen, oxygen, sulphur, selenium, or phosphorus. When the electron deficient atom boron is substituted into the ring and attached to two nitrogens with lone electron pairs, flattening of the ring results as seen in the case of 1,3-dimethyl-2-chloro-diazboracyclohexane[346]. This is probably caused by electron delocalization[346].

Miscellaneous six-membered rings. Electron diffraction studies have been made for several molecules which have no carbon atoms in their six-membered rings. Hexamethylcyclotrisiloxane (*74*)[348] is essentially planar with D_{3h} symmetry, while

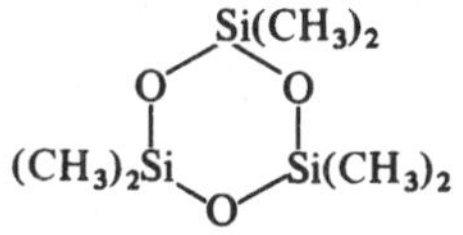

(*74*)

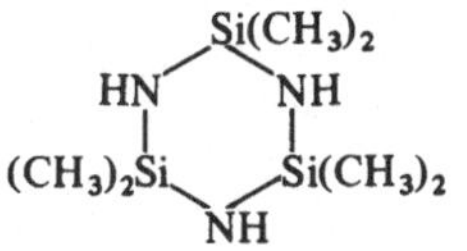

(*75*)

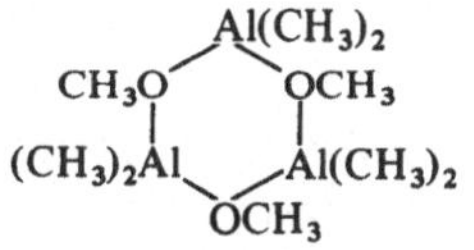

(*76*)

hexamethylcyclotrisilasane (*75*)[439] is puckered but the deviation from planarity is relatively small. The ring of methyl aluminium methoxy trimer (*76*)[350] is definitely non-planar. C_{3v} symmetry was assumed for this molecule, but lower symmetries could not be ruled out.

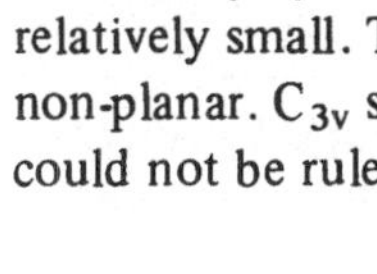

(*77*)

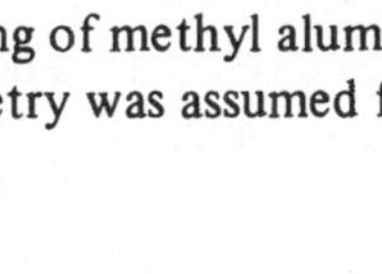

(*78*)

A chair conformation[351)] is suggested for Se_6 studied at 450 °C. $(MoO_3)_3$ (*77*) best fits a planar D_{3h} model[352)], and $(WO_3)_3$ (*78*) is perhaps puckered with C_{3v} symmetry[353)]. The former of these two molecules was studied at about 1000 °C and the latter at 1400 °C and shrinkage is thus probably quite important[2].

8.4 Rings with More than Six Atoms

The study of large rings by electron diffractions is generally complicated because of the large number of parameters to be determined. Often these rings have little symmetry or they may be quite flexible so that several conformations coexist. This, of course, adds to the already difficult problem. Quite often the only conclusion that can be made is that no one simple conformation can explain the experiment satisfactorily. Despite these obstacles, considerable progress has been made towards an understanding of structure and dynamics of several large rings by the means of electron diffraction. Molecular mechanics calculations have been made for many rings[27, 28, 285, 354)] with results which are often in good agreement with the experimental ones. Interconversion mechnisms of small, medium and large rings have been reviewed by Dale[355)]. Dunitz[356)] has reviewed recent X-ray work on medium size rings.

Seven-membered rings. Relatively few seven-membered rings have been studied in the free state. 1,3-cycloheptadiene (*79*)[357)] is planar except for C_6 which is bent

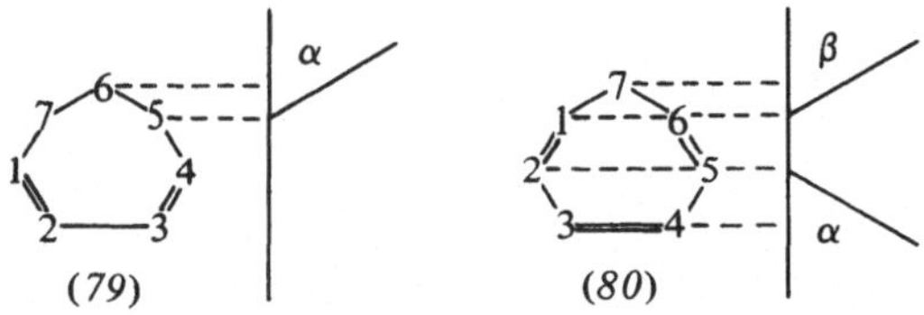

(*79*) (*80*)

63.9 ± 4.0 ° out of the plane formed by C_7, C_1, C_3, C_4, and C_5. 1,3,5-Cycloheptatriene (*80*) is non-planar with $\alpha = 40.5 \pm 2.0$ °, and $\beta = 36.5 \pm 2.0$ °. 2,4,6-Cycloheptatrien-1-one, tropone[359)], is assumed to be planar.

Eight-membered and larger rings. Cyclooctane[360)] was best accounted for by assuming a mixture of several conformations, since the ring is very flexible. *Trans*-cyclooctene[361)], cyclooctyne[362)], as well as 3,3,6,6-tetramethyl-1-thiacycloheptyne[363)] all prefer twisted conformations. 1,3-Cyclooctadiene (*81*)[364)] has C_i symmetry with a 37.8 ° angle between the two planar ethyelene groups. Cyclooctatetraene (*82*)[365)] has a tub form with $\alpha = 43.1 \pm 1.0$ °. *Cis,cis*-cyclodeca-1,6-diene

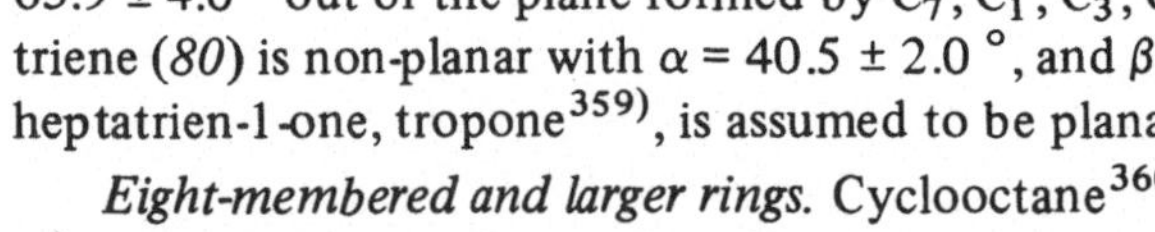

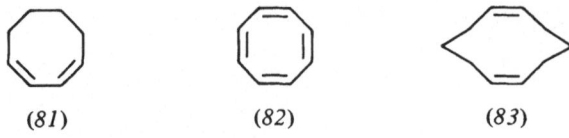

(*81*) (*82*) (*83*)

2 See Note Added in Proof.

(*83*)[366] has a predominant C_{2h} chair conformation. The situation encountered for cyclodecane[367] is quite similar to that found for cyclooctane[360] in that the ring is very flexible. A mixture of four different conformations yielded a good fit to the experimental data[367]. Cyclotretadeca-1,8-diyne is also a very flexible molecule[360].

Cope rearrangement is known to take place in semibullvalene (*84*)[368] and bullvalene (*85*)[369]. This process is quite slow compared to the electron-diffraction process, and the bond distances of (*84*) and (*85*) are therefore found to be similar to normal single and double bonds.

Miscellaneous medium-size rings. A few electron diffraction investigations have been made for medium size rings not containing carbon atoms in the ring. The eight-membered ring prosiloxane tetramer, $(H_2SiO)_4$[370], is best accounted for by assuming a puckered structure with S_4 symmetry. Cyclic tetrameric structures of lower symmetry cannot be ruled out. The situation in octamethylcyclotetrasiloxane[349], $[(CH_3)_2SiO]_4$, is similar. No well-defined conformation was found in the ten-membered ring decamethylcyclopentasiloxane[349], $[(CH_3)_2SiO]_5$, as well as in the twelve-membered ring dodecamethylcyclohexasiloxane[349], $[(CH_3)_2SiO]_6$, as a result of large amplitude vibrations.

Selenium trioxide[371] was studied at 120 °C. About 30% is monomeric SeO_3 and the rest is a tetrameric eight-membered ring presumably with S_4 symmetry. In dimethyl aluminium fluoride tetramer[372], $[AlF(CH_3)_2]_4$ the eight-membered ring consists of alternating aluminium and fluorine atoms. The molecule is probably non-rigid. A chair-boat model with C_s symmetry and two aluminium atoms in the mirror plane best fits experimental data. Further models with low symmetry are also possible.

(*84*)

(*85*)

9 Miscellaneous Large Amplitude Problems

Organometallic chemistry has in recent years been a rapidly expanding branch of chemistry. A large number of interesting and "unusual" or unexpected molecules have been synthesized. Electron-diffraction has been used to study many organometallic compounds. Several of these are quite flexible and exhibit challenging large amplitude problems. Recently, Haaland[373] has reviewed the electron-diffraction work on these molecules.

Bicyclopentadienylmetal and related compounds. Electron-diffraction has now been employed to study about ten bicyclopentadienylmetal compounds[373], $(C_5H_5)_2M$. With the notable exception of beryllium, the metal atom is always placed in the middle between the rings on the five-fold axis of symmetry. The structure obtained for $(C_5H_5)_2Be$ is shown in Fig. 21.

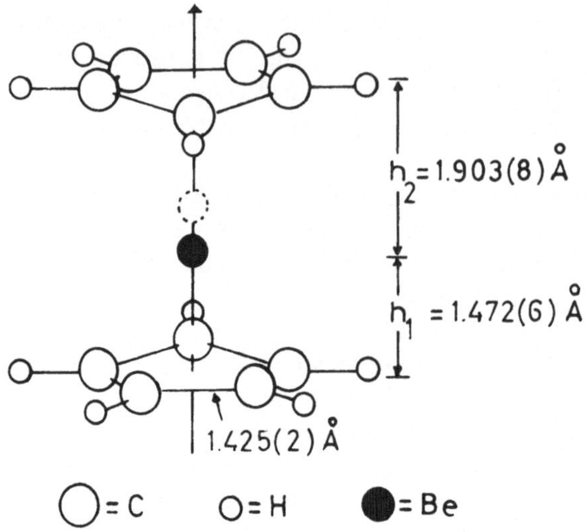

Fig. 21. Molecular structure of $(C_5H_5)_2Be$

In this case, the metal atom is closer to one ring than to the other. It has been suggested[374)] that the beryllium atom oscillates in a double minimum potential which has its minima on the five-fold axis of symmetry 22 pm on either side of the midpoint between the rings. However, theoretical calculations[375)] have failed to reproduce the experimental findings. A new electron-diffraction study of this molecule is now being made[376)]. Preliminary results show that the model of Fig. 21 is in agreement with the new data[376)]. Other models are also being studied[3].

The barriers restricting the torsional motion of the rings are definitely fairly low in all $(C_5H_5)_2M$ compounds which have hitherto been investigated[373)]. In one case, e. g. ferrocene, $(C_5H_5)_2Fe$ a barrier of 3.8 ± 1.3 kJ/mol has been determined by electron diffraction[22)]. Unfortunately, there are no other gas phase quantitative barrier determinations for the ring torsional motion of these "sandwich" compounds.

In a related compound, benzenechromium tricarbonyl, $C_6H_6Cr(CO)_3$[377)], internal rotation was seen to be nearly unhindered.

Molecules with several XF_3 groups. Some molecules of this type have been studied by electron-diffraction in recent years. In nickel tetrakistrifluorophosphine, $Ni(PF_3)_4$[378)], the PF_3 groups rotate nearly freely. The same was found for platinum tetrakisfluorophosphine, $Pt(PF_3)_4$[379)].

Two molecules presenting the same kind of torsional problems, are $C(CF_3)_4$ and $Ge(CF_3)_4$[380)]. The C-F distance is found to be nearly the same for the two compounds, but of course the distance from the central atom to the CF_3 carbon is considerably larger for the latter molecule. The C-C distance is found to be 156.2 pm, and the Ge-C distance is found to be 198.9 pm. This leads to a noticeable difference in the torsional motion of the two molecules. In $C(CF_3)_4$ the torsional barrier is estimated to be about 10 kJ/mol from the electron-diffraction study, while a CNDO/2 calculation estimated a barrier of about 3 kJ/mol. For $Ge(CF_3)_4$ the electron-diffraction data suggest free rotation. For these molecules the geared motion has not been considered as was the case for $(CF_3)_3CH$, previously discussed (p. 136).

3 See Note Added in Proof.

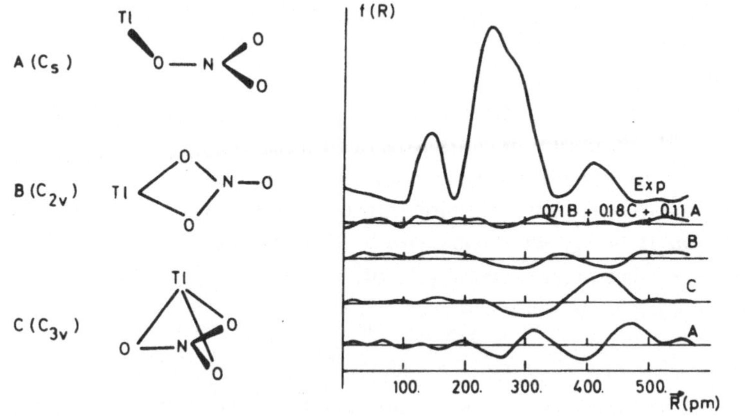

Fig. 22. Experimental and difference RD-curves (right) of various molecular models (left) of $TlNO_3$

High temperature studies of salts. Ionic compounds have in recent years been studied at elevated temperatures, especially in the Soviet Union. These distinguished studies have, in combination with other methods, demonstrated that polyatomic molecular species are rather the rule than the exception in the gaseous state at high temperatures. Moreover, large amplitude motions reflecting rather peculiar dynamics have been found to be of considerable importance in several of these investigations. In one such study, the one on thallium nitrate, $TlNO_3$[381], performed at 300–350 °C, the three models shown in Fig. 22 were considered. The root-mean-square amplitudes of vibration of the Tl-O distance was found to be in the 16–25 pm range for the various models A, B, or C. This is a very high value for two atoms linked to each other. None of the three models A, B or C was satisfactory. A composite model with a mole fraction of 0.71B, 0.18C, and 0.11A yielded a much better fit to the data as shown in Fig. 22. The exact nature of the unusual large amplitude motion observed in this case will need further investigation. There is also some evidence from other nitrates, e. g. $Cu(NO_3)_2$[382], $LiNO_3$[383], and $NaNO_3$[383] for a similar large amplitude behavior.

$TlReO_4$ also exhibits large amplitude motion[384]. It was suggested[384] that the thallium atom "orbits" around the near-spherical surface of the ReO-tetrahedron.

Tl_2SO_4, Cs_2WO_4, Cs_2SO_4 and Cs_2MoO_4 all[385] possess a structure which is presumably close to the D_{2d} structure (*86*). The extraordinary large root-mean-square amplitudes of vibration observed for these four compounds have been inter-

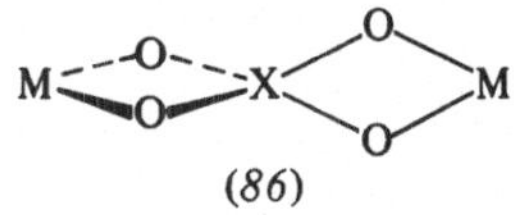

(*86*)

preted in a similar manner as for $TlReO_4$, namely as great displacements of the metal ions on the surface of an imaginary sphere formed by the acid residues. It is not unlikely that Tl_2MoO_4[386], $KAlCl_4$[387], $NaAlF_4$[388], $KYCl_4$[389], $TlInCl_4$[390], K_2CrO_4[391], In_2MoO_4[392], and K_2SO_4[391] behave similarly.

A theoretical discussion of the chemical bonding in molecules of this kind has recently been given[393].

Several alkali halides have been studied at elevated temperatures[394] and dimeric species forming four-membered rings have been found to predominate. There are indications of rather large amplitude vibrations in these dimers. It should be pointed out that monomeric species are also present in addition to the predominating dimers and several of these have been indentified by microwave spectroscopy[44].

Penta-, hexa-, and heptahalogenated compounds. Many compounds of the general type XY_n where Y is a halogen, mainly fluorine, and $n = 5,6$, or 7 have been investigated by electron diffraction in recent years. These substances exhibit a varied dynamical behavior. While WF_6[395, 396], MoF_6[395, 397], TeF_6[397, 398], UF_6[396, 397, 399], ReF_6[400], OsF_6[396], IrF_6[396], NpF_6[396], and PuF_6[396] are all rather "rigid" regular octahedrons of point group O_h, this is not the case for other similar molecules. PF_5[401], for example, is a trigonal bipyramide with axial PF bond length of 153.0 ± 0.4 pm and an equatorial PF bond length of 157.7 ± 0.5 pm. The NMR spectrum of this compound shows only one peak[401] which is indicative of a rather low barrier to pseudorotation whereby equatorial fluorine atoms are transformed into axial ones and *vice versa*. The AsF_5[402] and PCl_5[403] molecules behave in a similar manner. In CH_3PF_4[404] and $(CH)_2PF_3$[404] conformations with equatorial methyl groups are preferred. Very recently, $NbCl_5$[405] and $TaCl_5$[405] have been studied and found to have D_{3h} symmetries. Barriers to pseudototation were also determined as 6.0 ± 2.8 kJ/mol for $NbCl_5$ and 4.8 ± 2.4 kJ/mol for $TaCl_5$, respectively In XeF_6[406] effects of large amplitude vibrations are again manifest. No resolution of axial and equatorial XeF bond distances was possible. The molecular geometry is in the broad vicinity of C_{3v}. IF_7[407] and ReF_7[408] both show considerable distortion from a pentagonal bipyramid of D_{5h} symmetry. In IF_7[407] the average displacement of the equatorial fluorine atoms by ring puckering is 7.5 ° and the axial fluorine atoms are displaced by an average of 4.5 °. In ReF_7[408] the corresponding distortions are 8.7 ° and 7.5 °, respectively.

Torsional motion of nitromethanes. The nitro group of CH_3NO_2 undergoes nearly free rotation as shown by microwave spectroscopy[72]. Large amplitude torsional oscillation about the C-N bond also takes place in several other nitromethanes. In $C(NO_2)_4$[409], for example, the rotatory oscillations have an amplitude of 20 °. The mean amplitude of vibration for the interatomic distances depending on the torsion are as large as 13–17 pm at 45 °C. Similar results have been found for $CH(NO_2)_3$[410], and CH_2ClNO_2[411]. Interestingly, CCl_3NO_3 has a comparatively high barrier of about 12 kJ/mol[412]. Rather high barriers were also indicated for CBr_3NO_2[413] and CF_3NO_2[413]. This result is in disagreement with the microwave findings for the latter molecule, which was found to have nearly free rotation[75].

10 Concluding Remarks

The large amplitude motion that sometimes takes place within the molecule, should above all be studied in the gas phase. In a condensed phase intermolecular forces may in a systematic way hamper the internal motion. A large number of studies, taking advantage of a series of experimental techniques and theoretical methods, have for several decades been used to collect information in this field. In general a good qualitative picture of the large amplitude intramolecular motion has been formed, and sound theoretical systems have been developed for a quantitative approach to the problem. But in spite of this, a real quantitative description of the large amplitude motion has been given only for a few molecules.

The information collected so far is based upon findings of several methods. Unless a new thus far unknown method is developed, further success in this field depends on combination of various procedures among which gas electron diffraction is a prominent and useful one. Probably more systematically coordinated use of different experimental methods, in joint efforts to solve one and the same problem, is required to obtain more accurate and more detailed information.

Since the electron-diffraction method has been focused on in the present article, a question which naturally suggests itself is the following: Is there any special development in the field of electron diffraction that in particular may advance the study of large amplitude intramolecular motion? Since temperature enhances internal motion, the inclusion of high temperature study seems to be of increasing importance. In this field the Soviet electron-diffraction groups did the pioneering work, and they still lead the field. Combining high temperature studies with experiments done at the lowest possible temperatures would no doubt contribute favorably to our understanding of molecular flexibility.

We wish to conclude our essay with a final question: What is so important about the large amplitude motion to justify all the endeavor in describing it? A rationalist would base his answer on the general human wish to be useful and point to the applications that other fields of research may make of our findings. This can indeed be done with appreciable success. Generally, the flexibility of a molecule is described through its large amplitude potential, the determination of which is the main goal of our study. The interaction that takes place between molecules when they approach each other, must no doubt be dependent to a certain extent at least on the flexibility of the molecules involved. Accordingly the physical properties of a compound and its specific reactivity may also depend upon the molecular flexibility. It seems for example established beyond doubt that the existence of conformational option in a biologically active compound may be decisive for the specific properties of the compound.

But the question may also be answered differently. It is perhaps legitimate to claim that the fact that so many researchers are attracted and fascinated by the field and find it intellectually challenging, is a good enough justification in itself. For those who want to put their experimental or theoretical method to a critical test, large amplitude motion studies are to be recommended. Like many other fields of science, the study of intramolecular mobility is primarily carried out on its own merits, leaving possible application to future research.

Acknowledgements. The authors wish to express their gratitude to the Norwegian Research Council for Science and the Humanities for financial support of parts of the research reported in the present article. One of us (K. K.) also received a maintenance grant from the Council during the period when this article was prepared.

We are further indebted to those who helped us in preparing the manuscript. Mrs. Snefrid Gundersen made all the drawings, and the typing, re-typing and correction work was done by Miss Margareta Barth, Mr. Geir Baanrud, Mrs. Aslaug Jørgensen, and Mrs. Anne-Lise Ågren. We thank them all for their patience and many extra working hours.

11 References

1. Davisson, C. J., Germer, L. H.: Phys. Rev. *30*, 705 (1927)
2. Herzberg, G.: Molekülspektren und Molekülstrukturen. I. Zweiatomige Moleküle. Dresden und Leipzig: Verlag von Theodor Steinkopff 1939
3. Herzberg, G.: Molecular spectroscopy and molecular structure. II. Infrared and raman spectra of polyatomic molecules. New York, Cincinnati, Toronto, London, and Melbourne: Van Nostrand Reinhold Company 1945
4. Cyvin, S. J.: Molecular vibrations and mean square amplitudes. Oslo: Universitetsforlaget 1968
5. Bastiansen, O., Hassel, O.: Tidsskrift for Kjemi, Bergvesen og Metallurgi *8*, 96 (1946)
6. Bastiansen, O., Seip, H. M., Boggs, J. E.: Perspectives in Structural Chemistry *4*, 60 (1971)
7. Almenningen, A., Bastiansen, O., Fernholt, L., Hedberg, K.: Acta Chem. Scand. *25*, 1946 (1971)
8. Almenningen, A., Fernholt, L., Kveseth, K.: Acta Chem. Scand. *A 31*, 297 (1977)
9. Bastiansen, O., Hassel, O., Risberg, E.: Acta Chem. Scand. *9*, 232 (1955)
10. a) Seip, H. M., in: Molecular structure by diffraction methods, Vol. 1, p. 7. Sims, G. A. and Sutton, L. E. (eds.). London: The Chemical Society 1973
    b) Robiette, A. G.: ibid, p. 160, and references cited therein.
11. Bastiansen, O.: Acta Chem. Scand. *3*, 415 (1949)
12. a) Rundgren, J.: Arkiv för Fysik *35*, 269 (1967)
    b) Bastiansen, O., in: Structural chemistry and molecular biology, p. 640. Rich, A., and Davidson, N. (eds.). San Francisco and London: Freeman 1968
    c) Almenningen, A., Arnesen, S. P., Bastiansen, O., Seip, H. M., Seip, R.: Chem. Phys. Lett. *1*, 569 (1968)
    d) Ryan, R. R., Hedberg, K.: J. Chem. Phys. *50*, 4986 (1969)
    e) Gallaher, K. L., Bauer, S. H.: J. Chem. Phys. *78*, 2380 (1974)
13. a) Yates, A. C.: Computer Physics Commun. *2*, 175 (1971)
    b) Strand, T. G., Bonham, R. A.: J. Chem. Phys. *40*, 1686 (1964)
    c) Libermann, D., Walser, J. J., Cromer, D.: Phys. Rev. *137*, 1727 (1965)
14. a) Schäfer, L., Yates, A. C., Bonham, R. A.: J. Chem. Phys. *55*, 3055 (1971)
    b) Bonham, R. A., Schäfer, L.: International tables for X-ray crystallography IV. Birmingham: The Kynoch Press 1974
15. Kuchitsu, K., Cyvin, S. J.: Molecular structures and vibrations, p. 183. Cyvin, S. J. (ed.). Amsterdam: Elsevier 1972
16. a) Reitan, A.: Kgl. Norske Videnskabers Selskabs Skrifter, no. 2 (1958)
    b) Bartell, L. S.: J. Chem. Phys. *23*, 1219 (1955)
    c) Kuchitsu, K.: Bull. Chem. Soc. Jpn. *40*, 498 (1967)
17. a) Almenningen, A., Bastiansen, O., Munthe-Kaas, T.: Acta Chem. Scand. *10*, 261 (1956)
    b) Almenningen, A., Bastiansen, O., Traetteberg, M.: Acta Chem. Scand. *13*, 1699 (1959)
    c) Morino, Y.: Acta Cryst. *13*, 1107 (1960)
18. Morino, Y., Hirota, E.: J. Chem. Phys. *28*, 185 (1958)
19. Cyvin, S. J., Elvebredd, I., Cyvin, B. N., Brunvoll, J., Hagen, G.: Acta Chem. Scand. *21*, 2405 (1967)
20. Karle, J.: J. Chem. Phys. *45*, 4149 (1966)
21. Almenningen, A., Hartmann, Å. O., Seip, H. M.: Acta Chem. Scand. *22*, 1013 (1968)

22. Haaland, A., Nilsson, J. E.: Acta Chem. Scand. *22*, 2653 (1968)
23. Samdal, S.: Thesis, University of Oslo, 1973
24. Clark, A., Seip, H. M.: Chem. Phys. Lett. *6*, 452 (1970)
25. Thorson, W. R., Nakagawa, I.: J. Chem. Phys. *33*, 994 (1960)
26. Veillard, A.: Internal rotation in molecules, p. 404. Orville-Thomas, W. J. (ed.). London: Wiley 1974
27. Westheimer, F. H. in: Steric effects in organic chemistry, p. 523. Newman, M. S. (ed.). New York: Wiley 1956
28. Hendrickson, J. B.: J. Am. Chem. Soc. *83*, 4537 (1961) and J. Am. Chem. Soc. *89*, 7036,7047 (1967)
29. Chan, S. I., Ziem, J., Gwinn, W. D.: J. Chem. Phys. *44*, 1103 (1966)
30. Kilpatrick, J. E., Pitzer, K. S., Spitzer, R.: J. Am. Chem. Soc. *69*, 2483 (1947)
31. Buys, H. R., Altona, C., Havinga, E.: Tetrahedron *24*, 3019 (1968)
32. Meyer, R., Gammeter, A., Smith, P., Kühne, H., Nösberger, P., and Günthard, H. H.: J. Mol. Spectry. *46*, 397 (1973)
33. Nösberger, P., Bauder, A., Günthard, H. H.: Chem. Phys. *1*, 426 (1973)
34. Durig, J. R., Carter, R. O., Odom, J. D.: Inorg. Chem. *13*, 701 (1974)
35. Larsen, N. W., Mathier, E., Bauder, A., Günthard, H. H.: J. Mol. Spectry. *47*, 183 (1973)
36. Mathier, E., Welti, D., Bauder, A., Günthard, H. H.: J. Mol. Spectry. *37*, 63 (1971)
37. Ralowski, W.: Thesis, The Royal Institute of Technology, Stocholm 1975
38. Miller, F. A., Fately, W. G., Witkowski, R. E.: Spectrochim. Acta *23A*, 891 (1967)
39. Kakar, R. K.: J. Chem. Phys. *56*, 1189 (1971)
40. Hanyu, Y., Britt, C. O., Boggs, J. E.: J. Chem. Phys. *45*, 4725 (1966)
41. Høg, J. H., Nygaard, L., Sørensen, G. O.: J. Mol. Struct. *7*, 111 (1970)
42. Christen, D., Lister, D. G., Sheridan, J.: Chem. Soc. Trans. Faraday Soc. II *70*, 1953 (1974)
43. a) Lin, C. C., Swalen, J. D.: Rev. Modern Phys. *4*, 841 (1959)
    b) Dreizler, H.: Fortschritte d. Chem. Forsch. *10*, 59 (1968)
    c) Gordy, W., Cook, R. L.: Microwave molecular spectra. New York: Interscience Publisher 1970
44. a) Starck, B.: Molecular constants from microwave spectroscopy, Vol. 4. Landholt-Börnstein, New Series. Berlin, Heidelberg, New York: Springer 1967
    b) Demaison, J., Hüthner, W., Starck, B., Buck, J., Tischer, R., Winnewisser, M.: Molecular constants from microwave. Molecular beam, and electron spin resonance spectroscopy, Landolt-Börnstein, New Series, Vol. 7. Berlin, Heidelberg, New York: Springer 1974: Follow-up compilations available from Dr. B. Starck, University of Ulm, Germany
45. Weiss, S., Leroi, G. E.: J. Chem. Phys. *48*, 962 (1968)
46. Tuazon, E. C., Manocha, A. S., Fately, W. G.: Chem. Phys. Lett. *23*, 63 (1973)
47. Herschbach, D. R.: J. Chem. Phys. *25*, 358 (1956)
48. Durig, J. R., Bucy, W. E., Wurrey, C. J.: J. Chem. Phys. *63*, 5498 (1975)
49. Danti, A., Wood, J. L.: J. Chem. Phys. *30*, 582 (1959)
50. Tripton, A. B., Britt, C. O., Boggs, J. E.: J. Chem. Phys. *46*, 1606 (1967)
51. Eggers, D. F., Lord, R. C., Wickstrom, C. W.: J. Mol. Spectry. *59*, 63 (1976)
52. Schwendeman, R. H., Jacobs, G. D.: J. Chem. Phys. *36*, 1245 (1962)
53. Durig, J. R., Bucy, W. E., Wurrey, C. J.: J. Chem. Phys. *60*, 3293 (1974)
54. Allen, G., Brier, P. N., Lane, G.: Trans Faraday Soc. *63*, 824 (1967)
55. Lide, D. R.: J. Chem. Phys. *30*, 37 (1959)
56. Durig, J. R., Bucy, W. E., Carreira, L. A., Wurrey, C. J.: J. Chem. Phys. *60*, 1754 (1974)
57. Thomas, C. H., Nisbeth, K. D.: J. Chem. Phys. *61*, 5072 (1974)
58. Graner, G., Thomas, C. H.: J. Chem. Phys. *49*, 4160 (1968)
59. Brown, F. B., Clague, A. D. H., Heitkamp, N. D., Koster, D. F., Danti, A.: J. Mol. Spectry. *24*, 163 (1967)
60. Pauling, L.: The nature of the chemical bond. Ithaca, N. Y.: Cornell University Press 1960
61. Lees, R. M., Lovas, F. J., Kirchhoff, W. H., Johnson, D. R.: J. Chem. Phys. Reference Data *2*, 205 (1973)

62. Bauder, A., Günthard, H. H.: J. Mol. Spectry. *60*, 290 (1976)
63. Tagaki, K., Kojima, T.: J. Phys. Soc. Jpn. *30*, 1145 (1971)
64. Hirota, E.: J. Mol. Spectry. *43*, 36 (1972)
65. Rudolph, H. D., Trinkaus, A.: Z. Naturforsch. *23a*, 68 (1968)
66. Lees, R. M., Baker, J. G.: J. Chem. Phys. *48*, 5299 (1968)
67. Kreiner, W. A., Rudolph, H. D., Tan, B. T.: J. Mol. Spectry. *48*, 86 (1973)
68. Susskind, J.: J. Chem. Phys. *53*, 2492 (1970)
69. Woods, R. C.: J. Chem. Phys. *46*, 4789 (1967)
70. Turner, P. H., Cox, A. P.: Chem. Phys. Lett. *39*, 585 (1976)
71. Coffey, D., Britt, C. O., Boggs, J. E.: J. Chem. Phys. *49*, 591 (1968)
72. Rohart, F.: J. Mol. Spectry. *57*, 301 (1975)
73. Wilson, Jr., E. B., Naylor, Jr., R. E.: J. Chem. Phys. *26*, 1057 (1957)
74. Arnold, W., Dreizler, H., Rudolph, H. D.: Z. Naturforsch. *23a*, 301 (1968)
75. Tolles, W. M., Handelman, E. J., Gwinn, W. D.: J. Chem. Phys. *23*, 136 (1965)
76. Ogata, T., Cox, A. P.: J. Mol. Spectry. *61*, 265 (1976)
77. Caminati, W., Cazzoli, G., Mirri, A. M.: Chem. Phys. Lett. *35*, 475 (1975)
78. Ogata, T., Cox, A. P., Smith, D. L., Timms, P. L.: Chem. Phys. Lett. *26*, 186 (1974)
79. Cohan, E. A., Baudet, R. A.: J. Chem. Phys. *48*, 1220 (1968)
80. Meakin, P., Harris, D. O., Hirota, E.: J. Chem. Phys. *51*, 3775 (1969)
81. Munthe-Kaas, T.: Thesis, University of Oslo (1955)
82. Breed, H., Bastiansen, O., Almenningen, A.: Acta Cryst. *13*, 1108 (1960)
83. Almenningen, A., Bastiansen, O., Traetteberg, M.: Acta Chem. Scand. *15*, 1557 (1961)
84. Morino, Y., Iijima, T.: Bull. Chem. Soc. Jpn. *35*, 1661 (1962)
85. Cyvin, S. J.: Tidsskr. Kjemi, Bergvesen, Metallurgi *21*, 236 (1961)
86. Hargittai, I., Tremmel, J.: Coordination Chemistry Reviews *18*, 257 (1976)
87. Ezhov, Yu. S., Kasparov, V. V.: Seventh Austin Symposium on Gas Phase Molecular Structure, Austin, p. 58. Texas 1978
88. a) Kuchitsu, K.: J. Chem. Phys. *49*, 4456 (1968)
 b) Kuchitsu, K.: Bull. Chem. Soc. Jpn. *40*, 498 (1967)
 c) Kuchitsu, K.: Bull. Chem. Soc. Jpn. *44*, 96 (1971)
89. Bartell, L. S.: J. Chem. Phys. *23*, 1219 (1955)
90. Wharton, L., Berg, R. A., Klemperer, W.: J. Chem. Phys. *39*, 2023 (1963)
91. Büchler, A., Stauffer, J. L., Klemperer, W.: J. Chem. Phys. *40*, 3471 (1964)
92. Büchler, A., Stauffer, J. L., Klemperer, W.: J. Am. Chem. Soc. *86*, 4544 (1964)
93. Thompson, K. R., Carlson, K. D.: J. Chem. Phys. *49*, 4379 (1968)
94. Miller, F. A., Lemmon, D. H., Witkowski, R. E.: Spectrochim. Acta. *21*, 1709 (1965)
95. Smith, W. H., Leroi, G. E.: J. Chem. Phys. *45*, 1767 and 1784 (1966)
96. Brunvoll, J., Cyvin, S. J., Elvebredd, I., Hagen, G.: Chem. Phys. Lett. *1*, 566 (1968)
97. Tanimoto, M., Kuchitsu, K., Morino, Y.: Bull. Chem. Soc. Jpn. *43*, 2776 (1970)
98. Sabin, J. R., Kem, H.: J. Chem. Phys. *56*, 2195 (1972)
99. Weimann, L. J., Christoffersen, R. E.: J. Am. Chem. Soc. *95*, 2074 (1973)
100. Jensen, H. H., Nilssen, E. W., Seip, H. M.: Chem. Phys. Lett. *27*, 338 (1974)
101. Carreira, L. A., Carter, R. O., Durig, J. R., Lord, R. C., Milious, C. C.: J. Chem. Phys. *59*, 1028 (1973)
102. Duckett, J. A., Mills, I. M., Robiette, A. G.: J. Mol. Spectr. *63*, 249 (1976)
103. Weber, W. H., Ford, G. W.: J. Mol. Spectr. *63*, 445 (1976)
104. Glidewell, C., Robiette, A. G., Sheldrick, G. M.: Chem. Phys. Lett. *16*, 526 (1972)
105. Duckettt, J. A., Robiette, A. G., Mills, I. M.: J. Mol. Spectr. *62*, 34 (1976)
106. Barrow, M. J., Cradock, S., Ebsworth, E. A. V., Harding, M. M.: J. Chem. Soc. Chem. Comm. *1977*, 744
107. a) Orville-Thomas, W. J. (ed.): Internal rotation in molecules. London: Wiley 1974
 b) Robiette, A. G., in: Molecular structure by diffraction methods, Vol. 4, p. 45. Sims, G. A. and Sutton, L. E. (eds.). London: The Chemical Society 1976
 c) Bolm, R. K.; in: Molecular structure by diffraction methods, Vol. 5, p. 23 (1977)
108. Almenningen, A., Andersen, B., Traetteberg, M.: Acta Chem. Scand. *18*, 603 (1964)

109. Swick, D. A., Karle, I. L.: J. Chem. Phys. *23*, 1499 (1955)
110. Kveseth, K., Seip, H. M., Stølevik, R.: Acta Chem. Scand. *25*, 2975 (1971)
111. Elvebredd, I.: Acta Chem. Scand. *22*, 1606 (1968)
112. Almenningen, A., Bastiansen, O.: Acta Chem. Scand. *9*, 815 (1955)
113. Iijima, J.: Bull. Chem. Soc. Jpn. *46*, 2311 (1973)
114. Bartell, L. S., Higginbothan, H. K.: J. Chem. Phys. *42*, 851 (1965)
115. Gallaher, K. L., Yokozeki, A., Bauer, S. H.: J. Phys. Chem. *21*, 2389 (1974)
116. Andreassen, A. L., Bauer, S. H.: J. Chem. Phys. *56*, 3802 (1972)
117. Allen, G., Fewster, S.: Internal rotation in molecules, p. 268. Orville-Thomas, W. J. (ed.). London: Wiley 1974
118. Ivey, R. C., Schulze, P. D., Leggett, T. L., Kohl, D. A.: J. Chem. Phys. *60*, 3174 (1974)
119. Bartell, L. S., Boats, T. L.: J. Mol. Struct. *32*, 379 (1976)
120. Mann, D. E., Fano, L.: J. Chem. Phys. *26*, 1665 (1957)
121. Brunvoll, J., Hargittai, I., Seip, R.: Z. Naturforsch. *33a*, 222 (1978)
122. a) Vajda, E., Hargittai, I., Maltsev, A. K., Nefedov, O. M.: J. Mol. Struct. *23*, 417 (1974)
b) Hargittai, I., Brunvoll, J.: J. Mol. Struct. *44*, 107 (1978)
123. Beagley, B., Conrad, A. R., Freeman, J. M., Monaghan, J. J., Norton, B. G., Holywell, G. C.: J. Mol. Struct. *11*, 371 (1972)
124. Pfeiffer, M., Spangenberg, H. J.: Z. Phys. Chem. *232*, 47 (1966)
125. Oberhammer, H.: J. Mol. Struct. *31*, 237 (1976)
126. Rankin, D. W. H., Robertson, A.: J. Mol. Struct. *27*, 438 (1975)
127. Haase, J.: Z. Naturforsch. *28a*, 542 (1973)
128. Beagley, B., Managhan, J. J., Hewitt, T. O.: J. Mol. Struct. *8*, 401 (1971)
129. Bohn, R. K., Haaland, A.: J. Organometal. Chem. *5*, 470 (1966)
130. Hohn, C. H., Ibers, J. A.: J. Chem. Phys. *30*, 885 (1959)
131. Tanimoto, M., Kuchitsu, K., Morino, Y.: Bull. Chem. Soc. Jpn. *42*, 2519 (1969)
132. Chang, C. H., Andreassen, A. L., Bauer, S. H.: J. Org. Chem. *36*, 920 (1971)
133. Brunvoll, J.: Thesis, Trondheim 1962
134. van Schaick, E. J. M., Geise, H. J., Mijlhoff, F. L., Renes, G.: J. Mol. Struct. *16*, 23 (1973)
135. Harris, W. C., Holtzclaw, J. R., Kalasinsky, V. F.: J. Chem. Phys. *67*, 3330 (1977)
136. a) Kveseth, K.: Acta Chem. Scand. *A 28*, 482 (1974)
b) Kveseth, K.: Acta Chem. Scand. *A 29*, 307 (1975)
137. Allen, G., Fewster, S. in: Internal rotation in molecules, p. 279. Orville-Thomas, W. J. (ed.). London: Wiley 1974
138. Fernholt, L., Kveseth, K.: Acta Chem. Scand. *A 32*, 63 (1978)
139. Fujiyama, T., Kakimoto, M.: Bull. Chem. Soc. Jpn. *49*, 2346 (1976)
140. Wyn-Jones, E., Orville-Thomas, W. J.: Trans-Faraday Soc. *64*, 2907 (1968)
141. Bradford, W. F., Fitzwater, S., Bartell, L. S.: J. Mol. Struct. *38*, 185 (1977)
142. Sheppard, N., Szusz, G. J.: J. Chem. Phys. *17*, 86 (1949)
143. Fernholt, L., Kveseth, K.: Acta Chem. Scand., in press
144. Fitzgerald, W. E., Janz, J. G.: J. Mol. Spectrosc. *1*, 49 (1957)
145. Traetteberg, M., Bakken, P., Seip, R., Cyvin, S. J., Hopf, H.: J. Mol. Struct. *51*, 77 (1979)
146. Brown, D. E., Beagley, B.: J. Mol. Struct. *38*, 167 (1977)
147. Klaeboe, P., Nielsen, J. R.: J. Chem. Phys. *32*, 899 (1960)
148. Huisman, P., Mijlhoff, F. C.: J. Mol. Struct. *21*, 23 (1974)
149. Morino, Y., Kuchitsu, K.: J. Chem. Phys. *28*, 175 (1958)
150. Komaki, C., Ichishima, I., Kuratani, K., Miyazawa, T., Shimanouchi, T., Mizushima, S.: Bull. Chem. Soc. Jpn. *28*, 330 (1955)
151. Grindheim, S., Stølevik, R.: Acta Chem. Scand. *A 30*, 625 (1976)
152. Farup, P. E., Stølevik, R.: Acta Chem. Scand. *A 28*, 680 (1974)
153. Farup, P. E., Stølevik, R.: Acta Chem. Scand. *A 28*, 871 (1974)
154. Stølevik, R.: Acta Chem. Scand. *A 28*, 299 (1974)
155. Pauli, G. H., Momany, F. A., Bonham, R. A.: J. Am. Chem. Soc. *86*, 1286 (1964)
156. Ukaji, T., Bonham, R. A.: J. Am. Chem. Soc. *84*, 3631 (1962)

157. Pentin, Yu. A., Melikhova, L. P., Vl'Yanov, O. D.: Zh. Strukt. Khim. *4*, 5351 (1963)
158. Momany, F. A., Bonham, R. A., McCoy, W. H.: J. Am. Chem. Soc. *85*, 3077 (1963)
159. Kveseth, K.: to be published
160. Hedberg, K., Samdal, S.: private communication
161. Morino, Y., Miyagawa, I., Chiba, T., Shimozawa, T.: Bull. Chem. Soc. Jpn. *30*, 222 (1957)
162. Kuchitsu, K.: Bull. Chem. Soc. Jpn. *30*, 399 (1957)
163. Almenningen, A., Bastiansen, O., Harshbarger, F.: Acta Chem. Scand. *11*, 1059 (1957)
164. Ellestad, O. H., Kveseth, K.: J. Mol. Struct. *25*, 175 (1975)
165. Kveseth, K.: to be published
166. Kveseth, K.: Acta Chem. Scand. *A 32*, 51 (1978)
167. Abraham, R. J., Stølevik, R.: private communication
168. Stølevik, R.: Acta Chem. Scand. *A 31*, 359 (1977)
169. a) Bastiansen, O.: Om noen av de forhold som hindrer den fri dreibarhet om en enkeltbinding. Bergen: A. Garnaes boktrykkeri 1948
b) Almenningen, A., Bastiansen, O., Traetteberg, M.: Acta Chem. Scand. *12*, 1221 (1958)
c) Haugen, W., Traetteberg, M.: Selected topics in structural chemistry, p. 113. Andersen, P., Bastiansen, O., Furberg, S. (eds.). Oslo: Universitetsforlaget 1967
d) Hagen, K., Traetteberg, M.: private communication
e) Kveseth, K.: to be published
170. Lide, D. R. Jr., Jen, M.: J. Chem. Phys. *40*, 252 (1964)
171. Fateley, W. G., Harris, R. K., Miller, F. A., Witkowski, R. E.: Spectrochim. Acta *21*, 231 (1965)
172. Carreira, L. A.: J. Chem. Phys. *62*, 3851 (1975)
173. Lipnick, R. L., Garbisch, E. W. Jr.: J. Am. Chem. Soc. *95*, 6370 (1973)
174. Aston, J. G., Szasz, G., Wooley, H. W., Brickwedde, F. G.: J. Chem. Phys. *14*, 67 (1946)
175. Skaarup, S., Boggs, J. E., Skancke, P. N.: Tetrahedron *32*, 1179 (1976)
176. Skancke, P. N., Boggs, J. E.: J. Mol. Struct. *16*, 179 (1973)
177. Hagen, K., Hedberg, K.: The Norwegian Electron Diffraction Group, Annual Report 1977
178. Hagen, K.: The Norwegian Electron Diffraction Group, Annual Report 1977
179. Kuchitsu, K., Fukuyama, T., Morino, Y.: J. Mol. Struct. *1*, 463 (1967–68)
180. Hagen, K., Hedberg, K.: J. Am. Chem. Soc. *95*, 1003 (1973)
181. Hagen, K., Hedberg, K.: J. Am. Chem. Soc. *95*, 4796 (1973)
182. Hagen, K., Hedberg, K.: J. Am. Chem. Soc. *95*, 8266 (1973)
183. Hagen, K., Bondybey, V., Hedberg, K.: J. Am. Chem. Soc. *99*, 1365 (1977)
184. Vilkov, L. V., Sadova, N. I.: Zh. Strukt. Khim. *8*, 398 (1967)
185. Aten, C. F., Hedberg, L., Hedberg, K.: J. Am. Chem. Soc. *96*, 2463 (1968)
186. Traetteberg, M.: Acta Chem. Scand. *24*, 2295 (1970)
187. Chang, C. H., Andreassen, A. L., Bauer, S. H.: J. Org. Chem. *36*, 920 (1971)
188. Gundersen, G.: J. Am. Chem. Soc. *97*, 6342 (1975)
189. Rademacher, P.: Acta Chem. Scand. *26*, 1981 (1972)
190. a) Bastiansen, O., de Meijere, A.: Acta Chem. Scand. *20*, 516 (1966)
b) Hagen, K., Hagen, G., Traetteberg, M.: Acta Chem. Scand. *26*, 3649 (1972)
191. Stølevik, R., Schrumpf, G.: Acta Chem. Scand. *27*, 2694 (1973)
192. Stølevik, R., Schrumpf, G.: Acta Chem. Scand. *27*, 1950 (1973)
193. Smith, Z., Kohl, D. A.: J. Chem. Phys. *57*, 5448 (1972)
194. de Meijere, A., Lüttke, W.: Tetrahedron *25*, 2047 (1969)
195. Bartell, L. S., Guillory, J. P.: J. Chem. Phys. *43*, 647 (1965)
196. Bartell, L. S., Guillory, J. P., Parks, A. T.: J. Chem. Phys. *69*, 3043 (1954)
197. Bartell, L. S., Kohl, D. A.: J. Chem. Phys. *39*, 3097 (1963)
198. Fitzwater, S., Bartell, L. S.: J. Am. Chem. Soc. *98*, 8338 (1976)
199. Astrup, E. E.: Acta Chem. Scand. *A 31*, 125 (1977)
200. Grindheim, S., Stølevik, R.: Acta Chem. Scand. *A 31*, 69 (1977)
201. Johnsen, J. P., Stølevik, R.: Acta Chem. Scand. *A 29*, 457 (1975)

202. Fernholt, L., Stølevik, R.: Acta Chem. Scand. *A 29*, 651 (1975)
203. Johnsen, J. P., Stølevik, R.: Acta Chem. Scand. *A 29*, 201 (1975)
204. Fernholt, L., Stølevik, R.: Acta Chem. Scand. *A 28*, 963 (1974)
205. Stølevik, R., Thom, E.: Acta Chem. Scand. *25*, 3205 (1971)
206. Karle, J.: J. Chem. Phys. *59*, 3659 (1973)
207. Hoyland, J. R.: J. Chem. Phys. *49*, 1908 (1969)
208. Bastiansen, O.: Acta Chem. Scand. *3*, 408 (1949)
209. Dahr, J.: Indian J. Phys. *7*, 43 (1932)
210. Saunder, D. H.: Proc. Roy. Soc. (London) *A188*, 31 (1947)
211. Toussaint, J.: Acta Cryst. *1*, 43 (1948)
212. van Niekerk, J. N., Saunder, D. H.: Acta Cryst. *1*, 44 (1948)
213. Trotter, J.: Acta Cryst. *14*, 1135 (1961)
214. Hargreaves, A., Rizvi, S. H.: Acta Cryst. *15*, 365 (1962)
215. Almenningen, A., Bastiansen, O.: Kgl. Norske Vidensk. Selsk. Skr. *1958*, No 4
216. Bastiansen, O., Skancke, A.: Acta Chem. Scand. *21*, 587 (1967)
217. Friedman, P. S., Kopelman, R., Prasad, P. N.: Chem. Phys. Lett. *24*, 15 (1974)
218. Brenner, H. C., Hutchison, C. A., Kempler, M. D.: J. Chem. Phys. *60*, 2180 (1974)
219. Charbonneau, G., Delugeard, Y.: Acta Cryst. *B32*, 1420 (1976)
220. Casalone, G., Mariani, C., Mugnoli, A., Simonetta, M.: Acta Cryst. *B25*, 1741 (1969)
221. Brock, C. P., Kuo, M. S., Levy, H. A.: Acta Cryst. *B34*, 981 (1978)
222. Bastiansen, O.: Acta Chem. Scand. *6*, 205 (1952)
223. Farag, M. S.: Acta Cryst. *7*, 117 (1954)
224. Almenningen, A., Bastiansen, O., Skancke, P. N.: Acta Chem. Scand. *12*, 1215 (1958)
225. Bart, J. C. J.: Acta Cryst. *B24*, 1277 (1968)
226. Fischer-Hjalmars, I.: Tetrahedron *19*, 1805 (1963)
227. Golebriewski, A., Parczewski, A.: Theoret. Chim. Acta, *7*, 171 (1967)
228. Gondo, Y.: J. Chem. Phys. *41*, 3928 (1964)
229. Almlöf, J.: Chem. Phys. *6*, 135 (1974)
230. Fernholt, L., Rømming, C.: Private communication
231. Bastiansen, O., Fernholt, L.: Unpublished results
232. Almenningen, A., Bastiansen, O., Svendsås, P.: Acta Chem. Scand. *12*, 1671 (1958)
233. Skancke, A.: Acta Chem. Scand. *24*, 1389 (1970)
234. Bastiansen, O., Smedvik, L.: Acta Chem. Scand. *8*, 1593 (1954)
235. Bastiansen, O.: Acta Chem. Scand. *4*, 926 (1950)
236. Rømming, C., Seip, H. M., Aanesen Øymo, I.-M.: Acta Chem. Scand. *A28*, 507 (1974)
237. Hamor, M. J., Hamor, T. A.: Acta Cryst. *B34*, 863 (1978)
238. Ottersen, T.: Acta Chem. Scand. *A31*, 480 (1977)
239. Gwinn, W. D., Gaylord, A. S.: International review of science, physical chemistry series two, Vol. 3, p. 205. London, Boston: Butterworth 1976
240. Dunitz, J. D., Schomaker, V.: J. Chem. Phys. *20*, 1730 (1952)
241. a) Skancke, P. N.: Thesis, Norwegian Technical University, Trondheim (1960)
b) Almenningen, A., Bastiansen, O., Skancke, P. N.: Acta Chem. Scand. *15*, 711 (1961)
242. Ueda, T., Shimanouchi, T.: J. Chem. Phys. *49*, 470 (1968)
243. Kim, H., Gwinn, W. D.: J. Chem. Phys. *44*, 865 (1966)
244. a) Almenningen, A., Bastiansen, O., Walløe, L.: Tidskr. Kjemi, Bergv., Met. *25*, 206 (1965)
b) Almenningen, A., Bastiansen, O., Walløe, L.: Selected topics in structure chemistry, p. 91. Oslo: Universitetsforlaget 1967
245. Takabayashi, F., Kambara, H., Kuchitsu, K.: Seventh Austin Symposium on Gas Phase Molecular Structure, p. 63, 1978
246. Chang, C. H., Porter, R. F., Bauer, S. H.: J. Mol. Struct. *7*, 89 (1971)
247. Alekseev, N. V., Barzdain, P. P.: Zh. Struct. Khim. *15*, 181, (Eng. 171) (1974)
248. Bohn, R. K., Tai, Y.-H.: J. Am. Chem. Soc. *92*, 6447 (1970)
249. Andersen, B., Srinivasan, R.: Acta Chem. Scand. *26*, 3468 (1972)
250. Andersen, B., Fernholt, L.: Acta Chem. Scand. *24*, 445 (1970)

251. a) Mastryukov, V. S., Dorofeeva, O. V., Vilkov, L. V., Cyvin, B. N., Cyvin, S. J.: Zh. Strukt. Khim. *16*, 473 (Eng. 438) (1975)
b) Vilkov, L. V., Mastryukov, V. S., Oppenheim, V. D., Tarasenko, N. A.: Molecular structures and vibrations, p. 310. Amsterdam, London, New York: Elsevier Publishing Company 1972
252. Mastryukov, V. S., Dorofeeva, O. V., Vilkov, L. V., Hargittai, I.: J. Mol. Struct. *34*, 99 (1976)
253. Chan, S. J., Fernandez, J., Gwinn, W. D.: J. Chem. Phys. *33*, 1643 (1960)
254. Karakida, K., Kuchitsu, K.: Bull. Chem. Soc. Jpn. *48*, 1691 (1975)
255. Petit, M. G., Gibson, J. S., Harris, D. O.: J. Chem. Phys. *53*, 3408 (1970)
256. a) Smith, Z., Seip, R.: Acta Chem. Scand. *A30*, 759 (1976)
b) Chiang, J. F., Lu, K. C.: J. Phys. Chem. *81*, 1682 (1977)
257. Wehrung, T., Oberhammer, H., Haas, A., Koch, B.,.Welcman, N.: J. Mol. Struct. *35*, 253 (1976)
258. a) Wierl, R.: Ann. d. Ph. *8*, 521 (1931)
b) Wierl, R.: Ann. d. Ph. *13*, 453 (1932)
259. a) Hassel, O., Viervoll, H.: Tidsskr. Kjemi, Bergv. Met. *6*, 31 (1946)
b) Hassel, O., Viervoll, H.: Acta Chem. Scand. *1*, 149 (1947)
260. Adams, W. J., Geise, H. J., Bartell, L. S.: J. Am. Chem. Soc. *92*, 5013 (1970)
261. Höfler, F., Bauer, G., Hengge, E.: Spectrochim. Acta. *32 A*, 1435 (1976)
262. Smith, Z., Seip, H. M., Hengge, E., Bauer, G.: Acta Chem. Scand. *A 30*, 697 (1976)
263. a) Geise, H. J., Adams, W. J., Bartell, L. S.: Tetrahedron *25*, 3045 (1969)
b) Almenningen, A., Seip, H. M., Willadsen, T.: Acta Chem. Scand. *23*, 2748 (1969)
264. Smith, Z., Seip, H. M., Náhlovský, B., Kohl, D.: Acta Chem. Scand. *A 29*, 513 (1975)
265. Náhlovská, Z., Náhlovský, B., Seip, H. M.: Acta Chem. Scand. *23*, 3534 (1969)
266. Náhlovská, Z., Náhlovský, B., Seip, H. M.: Acta Chem. Scand. *24*, 1903 (1970)
267. Mamleev, A. H., Pozdeev, N. M., Magdesieva, N. N.: J. Mol. Struct. *33*, 211 (1976)
268. Geise, H. J., Mijlhoff, F. C.: Rec. Trav. Chim. *90*, 577 (1971)
269. Brendhaugen, K., Kolderup Fikke, M., Seip, H. M.: Acta Chem. Scand. *27*, 1101 (1973)
270. Almenningen, A., Fernholt, L., Rustad, S., Seip, H. M.: J. Mol. Struct. *30*, 291 (1976)
271. Almenningen, A., Fernholt, L., Seip, H. M.: J. Mol. Struct. *49*, 333 (1978)
272. Naumov, V. A., Semashko, V. N., Shaidulin, S. A.: Zh. Strukt. Khim. *14*, 595 (Eng. 555) (1973)
273. Almenningen, A., Kolsaker, P., Seip, H. H.: Acta Chem. Scand. *23*, 3398 (1969)
274. Kuczkowski, R. L., Gilles, C. W., Gallaher, K. L.: J. Mol. Spectry. *60*, 361 (1976)
275. Zaripov, N. M., Naumov, V. A.: Zh. Strukt. Khim. *14*, 588 (Eng. 551) (1973)
276. Schultz, G. Y., Hargittai, I., Martin, J., Robert, H. B.: Tetrahedron *30*, 2365 (1974)
277. Naumov, V. A., Semashko, V. N., Zav'yalov, A. P., Cherkasov, R. A., Grishina, L. N.: Zh. Strukt. Khim. *14*, 787 (Eng. 739) (1973)
278. Arbuzov, B. A., Naumov, V. A., Zaripov, N. M., Pronicheva, L. D.: Dokl. Akad. Nauk SSSR *195*, 1333 (Eng. 933) (1970)
279. Geise, H. J., van Laere, E.: Bull. Soc. Chim. Belg. *84*, 775 (1975)
280. Naumov, V. A., Pudovik, M. A.: Dokl. Akad. Nauk SSSR *203*, 351 (Eng. 237) (1972)
281. Naumov, V. A., Gulyaeva, N. A., Pudovik, M. A.: Dokl. Akad. Nauk SSSR *203*, 590 (Eng. 259) (1972)
282. Seip, H. M., Seip, R., Siebert, W.: Acta Chem. Scand. *27*, 15, (1973)
283. Almenningen, A., Seip, H. M., Vassbotn, P.: Acta Chem. Scand. *27*, 21 (1973)
284. Seip, H. M., Niedenzu, K.: J. Mol. Struct. *17*, 361 (1973)
285. 270 Ref. and references cited therein
286. Davis, M. I., Muecke, T. W.: J. Phys. Chem. *74*, 1104 (1970)
287. Chang, C. H., Bauer, S. H.: J. Phys. Chem. *75*, 1685 (1971)
288. Veniaminov, N. N., Alekseev, N. V., Bashkirova, S. A., Komalenkova, N. G., Chernyshev, E. A.: Zh. Strukt. Khim. *16*, 290 (Eng. 267) (1975)
289. Veniaminov, N. N., Alekseev, N. V., Bashkirova, S. A., Komalenkova, N. G., Chernyshev, E. A.: Zh. Strukt. Khim. *16*, 918 (Eng. 852) (1975)

290. Hilderbrandt, R. L., Peixoto, E. M. A.: J. Mol. Struct. *12*, 31 (1972)
291. Chang, C. H., Porter, R. F., Bauer, S. H.: Inorg. Chem. *8*, 1677 (1969)
292. Naumov, V. A., Semashko, V. N.: Zh. Strukt. Khim. *11*, 979 (Eng. 919) (1970)
293. Hilderbrandt, R. L., Wieser, J. D.: J. Mol. Struct. *22*, 247 (1974)
294. Rademacher, P.: J. Mol. Struct. *28*, 97 (1975)
295. Hassel, O.: Tidsskr. Kjemi, Bergv., Met. *3*, 32 (1943) Engl. transl. by Hedberg, K. in: Eliel, E. L., Allinger, N. L.: Topics in stereochemistry, Vol. 6. New York (1971)
296. a) Bastiansen, O., Fernholt, L., Seip, H. M., Kambara, H., Kuchitsu, K.: J. Mol. Struct. *18*, 163 (1973)
b) Ewbank, J. D., Kirsch, G., Schäfer, L.: J. Mol. Struct. *31*, 39 (1976)
297. Anet, F. A. L., Bourn, A. J. R.: J. Am. Chem. Soc. *89*, 760 (1967)
298. Allinger, N. L., Hickey, M. J., Kao, J.: J. Am. Chem. Soc. *98*, 2741 (1976)
299. a) Margrave, J. L., Frisch, M. A., Bautista, R. G., Clarke, R. L., Johnson, W. S.: J. Am. Chem. Soc. *85*, 546 (1963)
b) Allinger, N. L., Freiberg, L. A.: J. Am. Chem. Soc. *82*, 2393 (1960)
c) Johnson, W. S., Bauer, V. J., Margrave, J. L., Frisch, M. A., Dreyer, L. H., Hubbard, W. N.: J. Am. Chem. Soc. *83*, 606 (1961)
300. Cremer, D., Binkley, J. S., Pople, J. A.: J. Am. Chem. Soc. *98*, 6837 (1976)
301. Andersen, P.: Acta Chem. Scand. *16*, 2337 (1962)
302. a) Pierce, L., Nelson, R.: J. Am. Chem. Soc. *88*, 216 (1966)
b) Pierce, L., Beecher, J. F.: J. Am. Chem. Soc. *88*, 5406 (1966)
303. Aitkinson, V. A.: Acta Chem. Scand. *15*, 599 (1961)
304. Geise, H. J., Buys, H. R., Mijlhoff, F. C.: J. Mol. Struct. *9*, 447 (1971)
305. a) Geise, H. J., Mijlhoff, F. C.: J. Mol. Struct. *13*, 211 (1972)
b) Dallinga, G., Toneman, L. H.: Rec. Trav. Chim. *88*, 1221 (1969)
306. Askari, M., Schäfer, L.: J. Mol. Struct. *32*, 153 (1976)
307. a) Bastiansen, O., Hassel, O.: Nature *157*, 765 (1946)
b) Bastiansen, O., Hassel, O.: Tidsskr. Kjemi, Bergv., Met. *6*, 70 (1946)
308. Naumov, V. A., Bezzubov, V. M.: Dokl. Akad. Nauk SSSR *193*, 113 (Eng. 447) (1970)
309. Naumov, V. A., Bezzubov, V. M., Zaripov, N. M., Dashevskii, V. G.: Zh. Strukt. Khim. *11*, 801 (Eng. 743) (1970)
310. Naumov, V. A., Bezzubov, V. M.: Zh. Strukt. Khim. *8*, 530 (Eng. 446) (1967)
311. Braun, S., Traetteberg, M.: J. Mol. Struct. *39*, 101 (1977)
312. Naumov, V. A., Bezzubov, V. M.: Dokl. Akad. Nauk SSSR *186*, 599 (Eng. 408) (1969)
313. Spelbos, A., Mijlhoff, F. C., Bakker, V. H., Baden, R., van den Enden, L.: J. Mol. Struct. *38*, 155 (1977)
314. Schubert, W., Schäfer, L., Pauli, G. H.: J. Mol. Struct. *21*, 53 (1974)
315. Pauli, G. H., Askari, M., Schubert, W., Schäfer, L.: J. Mol. Struct. *32*, 146 (1976)
316. Askari, M., Pauli, G. H., Schubert, W., Schäfer, L.: J. Mol. Struct. *37*, 275 (1977)
317. Davis, M. I., Hassel, O.: Acta Chem. Scand. *18*, 813 (1964)
318. Naumov, V. A., Bezzubov, V. M.: Zh. Strukt. Khim. *13*, 977 (Eng. 914) (1972)
319. Aitkinson, V. A., Hassel, O.: Acta Chem. Scand. *13*, 1737 (1959)
320. Aitkinson, V. A., Lunde, K.: Acta Chem. Scand. *14*, 2139 (1960)
321. a) Schubert, W. K., Southern, T. F., Schäfer, L.: J. Mol. Struct. *16*, 403 (1973)
b) Haaland, A., Schäfer, L.: Acta Chem. Scand. *21*, 2474 (1967)
322. Bastiansen, O., Hassel, O.: Acta Chem. Scand. *5*, 1404 (1951)
323. Hjortaas, K. E., Strømme, K. O.: Acta Chem. Scand. *22*, 2965 (1968)
324. Hassel, O., Wang Lund, E.: Acta Cryst. *2*, 309 (1949)
325. Bastiansen, O., Ellefsen, Ø., Hassel, O.: Acta Chem. Scand. *3*, 918 (1949)
326. a) Chiang, J. F., Bauer, S. H.: J. Am. Chem. Soc. *91*, 1898 (1969)
b) Geise, H. J., Buys, H. R.: Rec. Trav. Chim. *89*, 1147 (1970)
c) Scharpen, L. H., Wollrab, J. E., Ames, D. P.: J. Chem. Phys. *49*, 2368 (1968)
327. Chiang, J. F., Lu, K. C.: Sixth Austin Symposium on Gas Phase Mol. Struct. *1976*, TA 6
328. a) Bastiansen, O.: Acta Chem. Scand. *6*, 875 (1952)
b) Bastiansen, O., Markali, J.: Acta Chem. Scand. *6*, 442 (1952)

329. a) Traetteberg, M.: Acta Chem. Scand. *22*, 2305 (1968)
b) Oberhammer, H., Bauer, S. H.: J. Am. Chem. Soc. *91*, 10 (1969)
c) Dallinga, G., Toneman, L. H.: J. Mol. Struct. *1*, 11 (1967)
330. Carreira, L. A., Carter, R. O., Durig, J. R.: J. Chem. Phys. *59*, 812 (1973)
331. Hagen, K., Traetteberg, M.: Acta Chem. Scand. *26*, 3636 (1972)
332. a) Schultz, Gy., Hargittai, I.: Acta Chim. Acad. Sci. Hung. *83*, 331 (1974)
b) Kewley, R.: Can. J. Chem. *50*, 1690 (1972)
333. Davis, M., Hassel, O.: Acta Chem. Scand. *17*, 1181 (1963)
334. Clark, A. H., Hewitt, T. G.: J. Mol. Struct. *9*, 33 (1971)
335. Astrup, E. E.: Acta Chem. Scand. *27*, 1345 (1973)
336. Adams, W. J., Bartell, L. S.: J. Mol. Struct. *37*, 261 (1977)
337. Schultz, G., Hargittai, I., Hermann, L.: J. Mol. Struct. *14*, 353 (1972)
338. Seip, R., Seip, H. M., Smith, Z.: J. Mol. Struct. *32*, 279 (1976)
339. Yokozeki, A., Kuchitsu, K.: Bull. Chem. Soc. Jpn. *44*, 2352 (1971)
340. Andersen, P., Astrup, E. E., Frederichsen, P. S., Nakken, K. F.: Acta Chem. Scand. *A 28*, 675 (1974)
341. Andersen, P., Astrup, E. E., Frederichsen, P. S., Nakken, K. F.: Acta Chem. Scand. *A 28*, 671 (1974)
342. Naumov, V. A., Zaripov, N. M., Shatrukov, L. F.: Zh. Strukt. Khim. *11*, 579 (Eng. 543) (1970)
343. Mustoe, F. J., Hencher, J. L.: Can. J. Chem. *50*, 3892 (1972)
344. Arbuzov, B. A., Naumov, V. A., Anonimova, I. V.: Dokl. Akad. Nauk SSSR *192*, 327 (Eng. 336) (1970)
345. Zaripov, N. M., Naumov, V. A.: Zh. Strukt. Khim. *14*, 588 (Eng. 551) (1973)
346. Seip, R., Seip, H. M.: J. Mol. Struct. *28*, 441 (1975)
347. Gallaher, K. L., Bauer, S. H.: J. Chem. Soc. Faraday Trans II *71*, 1173 (1975)
348. Oberhammer, H., Zeil, W., Fogarasi, G.: J. Mol. Struct. *18*, 309 (1973)
349. Rozsondai, B., Hargittai, I., Golubinskii, A. V., Vilkov, L. V., Mastryukov, V. S.: J. Mol. Struct. *28*, 339 (1975)
350. Drew, D. A., Haaland, A., Weidlein, J.: Z. Anorg. Allgem. Chem. *398*, 241 (1973)
351. Borzdain, P. P., Alekseev, N. V.: Zh. Strukt. Khim. *9*, 520 (Eng. 442) (1968)
352. Egorova, N. M., Rambidi, N. G.: Molecular structures and vibrations, p. 212. Cyvin, S. J. (ed.). Amsterdam: Elsevier 1972
353. Hargittai, I., Hargittai, M., Spiridonov, V. P., Erokhin, E. V.: J. Mol. Struct. *8*, 31 (1971)
354. a) Hendrickson, J. B.: J. Am. Chem. Soc. *86*, 4854 (1964)
b) Hendrickson, J. B.: J. Am. Chem. Soc. *89*, 7043 (1967)
c) Wiberg, K. B.: J. Am. Chem. Soc. *87*, 1070 (1965)
d) Allinger, N. L., Hirsch, J. A., Miller, M. A., Tyrminski, I., Van-Catledge, F. A.: J. Am. Chem. Soc. *90*, 1199 (1968)
e) Bixon, M., Lifson, S.: Tetrahedron *23*, 769 (1966)
f) Dunitz, J. D., Eser, H., Bixon, M., Lifson, S.: Helv. Chim. Acta *50*, 1512 (1967)
355. Dale, J.: Topics in stereochemistry, Vol. 9, p. 199. New York: Wiley & Sons 1976
356. Dunitz, J. D.: Perspectives of structural chemistry, Vol. II, p. 1. New York: Wiley, J. & Sons 1968
357. Hagen, K., Traetteberg, M.: Acta Chem. Scand. *26*, 3643 (1972)
358. Traetteberg, M.: J. Am. Chem. Soc. *86*, 4265 (1964)
359. Ogasawara, M., Iijima, T., Kimura, M.: Bull. Chem. Soc. Jpn. *45*, 3277 (1972)
360. Almenningen, A., Bastiansen, O., Jensen, H.: Acta Chem. Scand. *20*, 2689
361. Traetteberg, M.: Acta Chem. Scand. *B 29*, 29 (1975)
362. Haase, J., Krebs, A.: Z. Naturforsch. *26 a*, 1190 (1971)
363. Haase, J., Krebs, A.: Z. Naturforsch. *27 a*, 624 (1972)
364. Traetteberg, M.: Acta Chem. Scand. *24*, 2285 (1970)
365. Traetteberg, M.: Acta Chem. Scand. *20*, 1724 (1966)
366. Almenningen, A., Jacobsen, G. G., Seip, H. M.: Acta Chem. Scand. *23*, 1495 (1969)
367. Hilderbrandt, R. L., Wieser, J. D., Montgomery, L. K.: J. Am. Chem. Soc. *95*, 8598 (1973)

368. Wang, Y. C., Bauer, S. H.: J. Am. Chem. Soc. *94*, 5651 (1972)
369. Andersen, B., Marstrander, A.: Acta Chem. Scand. *25*, 1271 (1971)
370. Glidewell, C., Robiette, A. G., Sheldrick, G. M.: J. Chem. Soc. Chem. Comm. *1970*, 931
371. Mijlhoff, F. C.: Rec. Trav. Chim. *84*, 74 (1965)
372. Gundersen, G., Haugen, T., Haaland, A.: J. Chem. Soc. Chem. Comm. *1972*, 708
373. Haaland, A.: Topics Curr. Chem. *53*, 1 (1975)
374. a) Almenningen, A., Bastiansen, O., Haaland, A.: J. Chem. Phys. *40*, 3434 (1964)
b) Haaland, A.: Acta Chem. Scand. *22*, 3030 (1968)
375. a) Marynick, D. S.: J. Am. Chem. Soc. *99*, 1436 (1977)
b) Denar, M. J. S., Rzepa, H. S.: J. Am. Chem. Soc. *100*, 777 (1978)
c) Chiu, N.-S., Schäfer, L.: J. Am. Chem. Soc. *100*, 2604 (1978)
d) Jemmis, E. D., Alexandratos, S., Schleyer, P. v. R., Streitwieser, A., Schaefer, H. F.: J. Am. Chem. Soc. *100*, 5695 (1978)
e) Demuynck, J., Rohmer, M. M.: Chem. Phys. Lett. *54*, 567 (1978)
376. Almenningen, A., Haaland, A., Lusztyk, J.: Private communication
377. Chiu, N.-S., Schäfer, L., Seip, R.: J. Organomet. Chem. *101*, 331 (1975)
378. Almenningen, A., Andersen, A., Astrup, E. E.: Acta Chem. Scand. *24*, 1579 (1970)
379. Ritz, C. L., Bartell, L. S.: J. Mol. Struct. *31*, 73 (1976)
380. a) Oberhammer, H.: Seventh Austin Symposium on Gas Phase Molecular Structure 1978, p. 40, and private communication
b) J. Chem. Phys. *69*, 468 (1978)
381. Ischenko, A. A., Spiridonov, V. P., Zasorin, E. Z.: Zh. Strukt. Khim. *15*, 300 (Eng. 273) (1974)
382. La Villa, R. E., Bauer, S. H.: J. Am. Chem. Soc. *85*, 3597 (1963)
383. Khodchenkov, A. N., Spiridonov, V. P., Akishin, P. A.: Zh. Strukt. Khim. *6*, 765 (Eng. 724) (1965)
384. Roddatis, N. M., Tolmachev, S. M., Ugarov, V. V., Rambidi, N. G.: Zh. Strukt. Khim. *15*, 693 (Eng. 591) (1974)
385. Ugarov, V. V., Ezhov, Y. S., Rambidi, N. G.: J. Mol. Struct. *25*, 3571 (1975)
386. Tolmachev, S. M., Rambidi, N. G.: Zh. Strukt. Khim. *13*, 3 (Eng. 1) (1972)
387. Spiridonov, V. P., Erokhin, E. V., Lutoshkin, B. I.: Vestn., Moskv. Univ., Ser. II, Khim. *12*, 296 (1971)
388. Spiridonov, V. P., Erokhin, E. V.: Zh. Neorgan. Khim. *14*, 636 (Eng. 332) (1969)
389. Spiridonov, V. P., Brezgin, Y. A., Shakparanov, M. I.: Zh. Strukt. Khim. *12*, 1080 (Eng. 990) (1971)
390. Spiridonov, V. P., Brezgin, Y. A., Shakparanov, M. I.: Zh. Strukt. Khim. *13*, 320 (Eng. 293) (1972)
391. Spiridonov, V. P., Lutoshkin, B. I.: Vestn. Mosk. Univ. Ser. Khim. *25*, 509 (Eng. 1) (1970)
392. Tolmachev, S. M., Rambidi, N. G.: Zh. Strukt. Khim. *12*, 203 (Eng. 185) (1971)
393. a) Rambidi, N. G., J. Mol. Struct. *28*, 77 (1975)
b) Rambidi, N. G., J. Mol. Struct. *28*, 89 (1975)
394. Akishin, P. A., Rambidi, N. G.: Zh. Neorgan. Khim. *5*, 23 (Eng. 10) (1960)
395. Seip, H. M., Seip, R.: Acta Chem. Scand. *20*, 2698 (1966)
396. Kimura, M., Schomaker, V., Smith, D. W., Weinstock, B.: J. Chem. Phys. *48*, 4001 (1968)
397. a) Seip, H. M., in: Selected topics in structure chemistry, p. 25. Andersen, P., Bastiansen, O., Furberg, S. (eds.). Oslo: Universitetsforlaget 1967
b) Seip, H. M., Stølevik, R.: Acta Chem. Scand. *20*, 1535 (1966)
398. Gundersen, G., Hedberg, K., Strand, T. G.: J. Chem. Phys. *68*, 3548 (1978)
399. Seip, H. M.: Acta Chem. Scand. *19*, 1955 (1965)
400. Jacob, E. J., Bartell, L. S.: J. Chem. Phys. *53*, 2231 (1970)
401. Hansen, K. W., Bartell, L. S.: Inorg. Chem. *4*, 1775 (1965)
402. Clippard, Jr. F. B., Bartell, L. S.: Inorg. Chem. *9*, 805 (1970)
403. Adams, W. J., Bartell, L. S.: J. Mol. Struct. *8*, 23 (1971)
404. Bartell, L. S., Hansen, K. W.: Inorg. Chem. *4*, 1777 (1965)

405. Ischenko, A. A., Strand, T. G., Demidov, A. V., Spiridonov, V. P.: J. Mol. Struct. *43*, 227 (1978)
406. Bartell, L. S., Gavin, Jr. R. M.: J. Chem. Phys. *48*, 2466 (1968)
407. Adams, W. J., Thompson, H. B., Bartell, L. S.: J. Chem. Phys. *53*, 4040 (1970)
408. Jacob, E. J., Bartell, L. S.: J. Chem. Phys. *53*, 2235 (1970)
409. Sadova, N. I., Popik, N. I., Vilkov, L. V.: J. Mol. Struct. *31*, 399 (1976)
410. Sadova, N. I., Popik, N. I., Vilkov, L. V., Pankrushev, Y. A., Shlyapochnikov, V. A.: J. Chem. Soc. Chem. Comm. *1973*, 708
411. Sadova, N. I., Vilkov, L. V., Anfimova, J. M.: Zh. Strukt. Khim. *13*, 763 (Eng. 717) (1972)
412. Knudsen, R. E., George, C. F., Karle, J.: J. Chem. Phys. *44*, 2334 (1966)
413. Karle, I. L., Karle, J.: J. Chem. Phys. *36*, 1969 (1962)
414. Hargittai, I., Schulz, Gy., Naumov, V. A., Kitaw, Yu. P.: Acta Chim. Acad. Sci. Hung. *90*, 165 (1976)
415. Spiridonov, V. P., Ishchenko, A. A., Zasorin, E. Z.: Uspekhi Khimli *47*, 101 (1978)
416. Spiridonov, V. P., Ishchenko, A. A., Zasorin, E. Z.: Russian Chem. Rev. *47*, 1 (1978)
417. Lolck, J.-E., Brodersen, S.: J. Mol. Spectr. *72*, 445 (1978)
418. Cyvin, B. N., Hargittai, M., Cyvin, S. J., Hargittai, I.: Acta Chim. (Budapest) *84*, 55 (1975)
419. Almenningen, A., Haaland, A., Lusztyk, J.: J. Organometal. Chem., in press

Received June 28, 1978

Note Added in Proof

1. An important article written in Russian[415] on a similar subject as the present one, was unfortunately overlooked during the preparation of our manuscript. The article has since been published in English[416]. The main emphasis of the Russian article is on dynamic effects of inorganic molecules.
2. Since this manuscript was prepared a new value for the ν_7-frequency in C_3O_2 has been published[417].
3. Concerning the molecule $(WO_3)_3$ shrinkage has been calculated[418]. The inclusion of the shrinkage effect does not change the conclusion that the ring has a puckered conformation.
4. A slip sandwich model derived from the C_{5v} shown in Fig. 21 by moving the ring that is at the greatest distance from Be, sideways, while the two rings remain essentially parallel, is found to be in even better agreement with the electrondiffraction data than the C_{5v} model. It is likely that the far ring undergoes large amplitude motion in this direction, but it remains undecided whether the *equilibrium* structure is C_{5v} or not.

Molecular models of D_{5h} or D_{5d} symmetry or models containing one π-bonded and one σ-bonded ring are not in agreement with the ED data[419].

Author Index Volumes 26–81

The volume numbers are printed in italics